T0251974

Designing Embedded Communications Software

T. Sridhar

CRC Press
Taylor & Francis Group
Boca Raton London New York

CRC Press is an imprint of the
Taylor & Francis Group, an **informa** business

CRC Press
Taylor & Francis Group
6000 Broken Sound Parkway NW, Suite 300
Boca Raton, FL 33487-2742

First issued in hardback 2017

Copyright © 2003, by Taylor & Francis.
CRC Press is an imprint of Taylor & Francis Group, an Informa business

No claim to original U.S. Government works

ISBN 13: 978-1-138-41220-0 (hbk)
ISBN 13: 978-1-57820-125-9 (pbk)

This book contains information obtained from authentic and highly regarded sources. Reasonable efforts have been made to publish reliable data and information, but the author and publisher cannot assume responsibility for the validity of all materials or the consequences of their use. The authors and publishers have attempted to trace the copyright holders of all material reproduced in this publication and apologize to copyright holders if permission to publish in this form has not been obtained. If any copyright material has not been acknowledged please write and let us know so we may rectify in any future reprint.

Except as permitted under U.S. Copyright Law, no part of this book may be reprinted, reproduced, transmitted, or utilized in any form by any electronic, mechanical, or other means, now known or hereafter invented, including photocopying, microfilming, and recording, or in any information storage or retrieval system, without written permission from the publishers.

For permission to photocopy or use material electronically from this work, please access www.copyright.com (http://www.copyright.com/) or contact the Copyright Clearance Center, Inc. (CCC), 222 Rosewood Drive, Danvers, MA 01923, 978-750-8400. CCC is a not-for-profit organization that provides licenses and registration for a variety of users. For organizations that have been granted a photocopy license by the CCC, a separate system of payment has been arranged.

Trademark Notice: Product or corporate names may be trademarks or registered trademarks, and are used only for identification and explanation without intent to infringe.

**Visit the Taylor & Francis Web site at
http://www.taylorandfrancis.com**

**and the CRC Press Web site at
http://www.crcpress.com**

Cover layout design: Damien Castaneda

Table of Contents

Foreword

This document describes the usage and input syntax of the Unix Vax-11 assembler As. As is designed for assembling code produced by the 'C' compiler; certain concessions have been made to handle code written directly by people, but in general little sympathy has been extended.

— Berkeley Vax/Unix Assembler Reference Manual (1983)

Designing software for embedded communication systems is mostly mysterious for an ordinary software developer even though such systems have been designed, developed, and deployed for decades in many different environments. Typically, knowledge about the specific challenges and issues encountered in designing and building embedded communications software is known in terms of anecdotes and folklore among the developers of embedded systems. Conferences like Communications Design Conference have recently started elevating this topic to the center stage, but this specific area of software design continues to remain mysterious. In contrast, popularity of the Web and Web-based services over less than a decade has spawned a wealth of technical literature on software design and development for Web Services.

Design and development of communication systems has experienced a major trend in recent years. Increasing emphasis on reducing the R&D costs has led to adoption of both merchant silicon and merchant (or third-party) software for building embedded communication systems. This trend has wide-spread ramifications requiring a new breed of software developers that understand building and integrating a wide variety of reusable software components that come together in networking systems such as switches, routers, traffic aggregators (DSLAM, CMTS), and load balancers. A variety of vendors, currently offering network processors, co-processors, and software stacks or individual protocol suites, attests to this trend.

For this trend to scale up, formal knowledge of how to design and develop embedded communication software for networking systems is necessary for beginners as well as experienced professionals. I am delighted to see that T. Sridhar, with his extensive experience in product development, has stepped up to fill the void with this book.

This is a very well organized book. It serves the needs of both a novice and an experienced programmer. For example, it first starts with an overview of the OSI Reference model, role of protocol software components, device drivers, and a list of design considerations that must be taken into account in early stages of product design. It then systematically walks the reader through the considerations specific to the communications software design including issues related to partitioning of functionality in communications software.

Once the reader has grasped these concepts, Sridhar walks the reader through the details of data structure design, buffer and timer management which form the backbone of any communications software component and are important for achieving a high-performance product. I really enjoyed the chapter on multi-board development which addresses an often ignored but difficult part of software design in this area. Finally, the book introduces the reader to the different phases of software development process and finishes with real-life examples of communication software design used in two of the most popular commercial products.

At Intel, we are excited at the opportunity to transform the networking industry by providing programmable building blocks. A part of this goal is to build a strong ecosystem of communication software providers that allow integration of complete networking systems out of merchant software and silicon. This book makes a significant contribution by de-mystifying important aspects of communications software design. Curricula at many universities today lack a good introduction to communication software design. In addition, for professional developers looking to participate in the emerging ecosystem of communication software design and development, this book is a perfect starting point to understand the basic principles and issues encountered in this area. I strongly recommend this book for both types of audience.

Raj Yavatkar
Chief Software Architect
Network Processing Group
Intel Corporation
May 2003

Preface

With the rapid adoption of the Internet, communications devices have increased in importance. These devices are used in various parts of the network starting from network enabled PDAs and pagers, right up through to complex Central Office switches. Most of these network devices haves a robust communications software function, which is used to communicate with other devices as well as with a controlling entity like a network manager.

As in other areas of engineering, design is the first step in developing a product. There are several books on networking and communication including engineering approaches to network systems design. This book focuses primarily on the software aspect of communications systems – specifically those used to build embedded communications devices. Host systems have had protocol and networking functionality for several years – they are in fact, treated as a part of the OS. This book focuses on embedded communications systems, specifically those which use a real time operating system.

This book approaches communications software design from the perspective of a designer of embedded systems software. It assumes a knowledge of real time concepts including tasks, interrupts, scheduling and inter process communication.. It incorporates several issues from engineering folklore and best practices at various engineering organizations. Readers might be familiar with some of the concepts since they might have seen these addressed in their own internal company documentation or in some white papers from industry vendors.

During my years in designing and developing communications software, I have been fortunate to be associated with several communications software experts. These individuals have provided me with and pointed me to various tips and techniques for communications software design. This book grew out of a need to capture several of those issues, so that engineers venturing into communications software design have a good foundation.

Target Audience

The basic audience is embedded engineers who are writing communications software. This includes both people who are just venturing into communications software development as well as those who have some experience in the area. The first group of people will be able to obtain information in one place – information that they would normally have to glean from articles, colleagues, internal documents and some Web sites. The second group of people would obtain an idea of some of the other issues in the system – for example, protocol stack developers would learn about system architecture and software.

The audience also includes practicing embedded engineers who are just starting to write communications software, as well as graduate/undergraduate students working on communications software projects.

Organization of the Book

Chapter 1 discusses the OSI 7 Layer model in the context of software based implementations. It provides an overview of some of the issues in communications equipment, a foundation for building the software for these devices.

Chapter 2 details the various factors involved in software design for communications systems. It discusses host and embedded communications software requirements, including RTOS, protocol stack and hardware acceleration including design tradeoffs. Details of engineering software to work with and without hardware acceleration are also provided

Chapter 3 revisits layering in the context of software partitioning including why it is difficult to maintain strict layering. It outlines tasks, modules and their interface requirements.

Chapter 4 is a detailed description of protocol stacks and their implementation. State tables and their implementation, interfaces between protocol modules and management of protocol stacks are discussed in this chapter.

Chapter 5 provides the design issues with respect to tables used in communications software. Tables may be required for configuration, status and statistics and for protocol operation. Data structure allocation, design and access mechanisms for tables are outlined.

Chapter 6 provides a detailed view of buffer and timer management schemes in communications software design. It discusses the mbuf and STREAMS buffer schemes, discusses timer design including the use of a timer task and events.

Chapter 7 details management software design in communications systems. Management schemes, use of management protocol abstraction, saving and restoring configuration are some of the key issues covered in this chapter.

Chapter 8 is a discussion of issues with designing software for multi CPU and multi board systems. Popular multi board architectures, inter CPU messaging layer abstraction and redundancy are covered in this chapter.

Chapter 9 is a practitioner level view of communications software design and development. Details about the development phases and their outputs, hardware independent and COTS board testing are described in this chapter.

Acknowledgements

I am very thankful to my employer FutureSoft and its CEO Mr. K.V. Ramani for their support. I have been fortunate to be associated with several excellent engineers, who have helped in the ideas for this book (even if they did not know it!)– KK.Srinivasan, S. Ravikumar, Dr. Raj Yavatkar, Vijay Doraiswami, Kwok Kong, Manikantan Srinivasan , Rajesh Kumar, Elwin Eliazer and several others.

I wish to express my sincere thanks to Dr. Raj Yavatkar, Chief Software Architect at Intel, who agreed to write a foreword for the book.

Robert Ward of CMP Books was the person who worked with me at the beginning of this project to make this book a reality. Michelle O'Neal was a constant source of encouragement and worked wonders with her project management. Justin Fulmer was very patient with my edits and corrections. Paul Temme, who ably handled the entire project, kept everything running.

I will be failing in my duty if I did not mention Sue Thorstensen, the technical editor. She ensured that I did not stray, polished the manuscript in several areas and made it readable. Manikantan Srinivasan, Vijay Doraiswami and Sandhya Ravikumar provided valuable comments. Mani was very diligent in pointing out repetitions and inconsistencies, especially in the fine print.

This book would not have been possible without the support and encouragement of my wife Padmini Sridhar, who was behind me throughout this process. Ramasamy Rathnam and Asokan Selvaraj were two individuals who encouraged me to put my thoughts down on paper. I am also thankful to my support network which included Mrs. T. Saraswathi, Dr. Girija Suresh., R. Suresh, Sunder Mahalingam and Dr. Suresh Gopalan, among others. Suresh, in particular, kept me going when I was getting bogged down.

This book is intended to be a practitioner's guide to embedded communications software design. It would have served its purpose if it helps you understand and accelerate one or more phases of your next project. I take responsibility for any errors/omissions that may still remain in the text. I hope you will find this book useful.

Updates

Those of you who would like to sign up for e-mail news updates can send a blank e-mail to embeddedcoms@news.cmpbooks.com. If you have a suggestion or correction, please email your comments to: tsridhar@mail.com.

CHAPTER 1

Introduction

Communications systems include many devices ranging in complexity from small hand-held phones to large, central office switching devices. The earliest communications devices, such as phones, were all electrical and did not possess software, but new communications devices incorporate software based on the function that each device performs in the network. For example, a cellular phone has a microprocessor running a protocol stack to communicate with the cellular network. Frequently, it has additional capabilities, such as the ability to download software upgrades from the network or connect with the Internet.

Understanding issues common to communications equipment is the first step in developing a communications software strategy. Hardware variations then need to be considered in relation to these common issues. In some systems, the size of the code may play a more important role than performance, while in others, complete protocol functionality may not be required. Through a review of the Open Systems Interconnect (OSI) seven layer model, this chapter provides an introduction to the various types of communications systems and specifies a context for the software functions for each of these layers.

1.1 OSI Reference Model

The Open Systems Interconnect (OSI) model (see Table 1.1) was created by the International Organization for Standardization (ISO) to form the basis for communications systems. The OSI seven-layer model for communication protocols provides a modular separation of functionality into seven layers, which can be implemented in hardware and/or software. Each layer works independently, yet builds upon the lower ones.

The seven-layer model is useful for educational and comparative purposes, but most real-world implementations deviate somewhat to accommodate specific application requirements.

1

Each of the seven layers implements a specific communications function. This logical division allows for modular development, ease of upgrades, and increased manageability, as shown in Table 1.1.

Table 1.1   OSI reference model.

Layer	Typical Implementation	Name	Function	Examples
7	Software	Application	Network interface and user apps.	Email
6	Software	Presentation	Data format	XDR (eXtended Data Representation) in Network File System (NFS) protocol definition
5	Software	Session	Dialog and synchronization management	Checkpointing protocols
4	Software	Transport	End-to-End Application Transport	User Datagram Protocol (UDP), Transmission Control Protocol (TCP)
3	Software and Hardware	Network	Addressing and packet transmission	Internet Protocol (IP)
2	Hardware and Software	Data Link	Frame transmission across link	Ethernet MAC
1	Hardware	Physical	Transmission method	RS-232

Each layer is implemented via a combination of end points and protocols. An end point is a device that implements a protocol function. It communicates with a *peer* end point using the implemented protocol. For example, two hosts using the Transmission Control Protocol (TCP) to communicate between each other are TCP peers (Layer 5 peers, as depicted in Figure 1.1). Similarly, a router and a WAN switch communicating over a Frame Relay (FR) link are FR peers.

The following sections will discuss each of these layers, with specific emphasis on the software aspects.

1.1.1   Physical Layer

This layer defines how transmission media, such as cable, communicate with other devices and how bits are conveyed between peer systems. The physical layer provides the hardware means for transmitting data, including mechanical, procedural, electrical, and functional specifications for attributes such as voltage levels, encoding formats, and signal pins. This layer is almost always implemented in hardware; functionality for the physical layer is usually enabled in software as device drivers.

Figure 1.1 **OSI Reference Model and Communication between Peers.**

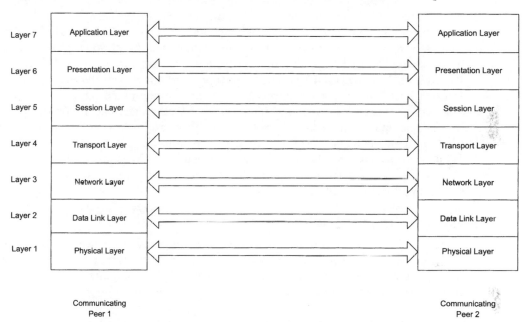

1.1.2 Data Link Layer

The physical layer is inherently unreliable—its susceptibility to electrical noise being one of the reasons. The data link layer provides for reliable transmission of frames between peer nodes. As with the physical layer, the peer node is a physically adjacent node. The data link layer is subdivided into two layers—the Logical Link Control (LLC) layer and the Media Access Control (MAC) layer. The data link layer is usually associated with data integrity through error detection and uses data checks such as Cyclic Redundancy Check (CRC). Framing and CRC are handled by hardware controllers; so the data link layer software needs to program the controllers appropriately, as with the physical layer.

Other functions may need to be implemented completely in software, with the hardware controller used only for the framing. For example, a data link layer protocol such as link accessin ISDN (Integrated Services Digital Network) networks is a Layer 2 protocol.

1.1.3   Network Layer

This layer is responsible for delivery of packets from the source to the destination. It isolates the network topology from the higher layers, so that these layers are network independent. The network technologies may be different—for example, the source may be connected to an Ethernet network, while the destination may be part of an ISDN network. The network layer provides information for source and destination addresses, subnet information, and parts used by higher transport layers. This layer uses information to *determine* the correct path to reach a destination from the source and proceeds with *forwarding* the packets from source to destination networks.

Routers and software used by routers typically utilize the network layer.

Internet Protocol (IP) is an example of a network layer protocol. It is a connectionless protocol, in the sense that individual packets (datagrams) can be treated differently—for example, each packet can potentially follow a different path from source to destination, depending upon the forwarding decision at each router.

While network layer forwarding can be implemented in either software or hardware, routing is usually less performance critical and involves more peer transactions. Therefore, routing is usually implemented in software.

1.1.4   Transport Layer

Running on top of the network layer, the transport layer provides network-independent, end-to-end integrity between two devices communicating through the network. It builds on the addressing protocols implemented in the network layer and interfaces with higher layer processes and applications on both source and destination systems. Protocols at this layer may be either connection oriented and reliable or connectionless and unreliable.

A connectionless transport protocol, such as User Datagram Protocol (UDP), does not provide feedback from the receiver and can be unreliable. A connection-oriented protocol, such as Transmission Control Protocol (TCP), provides feedback on the reception of the data by the peer. A connection process is initiated prior to data transmission, an acknowledgement is sent upon receipt of the data, and error detection and recovery routines ensure the data arrives intact. The connection is closed when the transmission is complete. For these capabilities, TCP is considered to be reliable.

Transport layer functions are usually implemented in software, except for architectures that support a large number of connections. In this situation, an off-CPU adapter or dedicated chip may be used for handling processing-intensive TCP functions, including data movement. Often TCP Offload Engines (TOE) are used in large-scale communications systems to move processing away from the server. To the higher layers, the TCP interface is preserved, but the actual implementation of TCP uses hardware.

1.1.5   Session, Presentation and Application Layers

The session, presentation, and application layers are closest to the user applications and can be treated together. Session layer functionality includes establishment, management, and termination of application connections and includes services such as data flow synchronization, partitioning, and checkpointing. The presentation layer specifies how user applications format data between applications and includes functionalities such as encryption, data compression, and character sets. The application layer provides end-user services such as mail, file transfer, and so on.

The TCP/IP world uses only one layer—the application layer—to signify all of these, even though the OSI model specifies each of them as separate layers. Hence, the TCP/IP model is a five-layer model. Protocols like Simple Mail Transfer Protocol (SMTP), File Transfer Protocol (FTP), and Virtual terminal protocol (Telnet) are in the application layer. These layers are usually implemented as networking applications on the communications system, but some functions, such as encryption algorithms, may run in an off-CPU hardware accelerator for security or performance reasons.

1.1.6 Networking Communication

Individually, each layer implements a specific communication function that is modular and independent of other layers. Network communication processes data from the bottom of the OSI model with peer devices in the same layer. For example, the physical layer of device A will communicate with the physical layer of device B. So is the case for the data link layer, but the data link layer has to use the services of the physical layer to communicate with the data link layer on device B. Attached headers and trailers encapsulate data and provide a communication path from layer to layer.

From an encapsulation perspective, the lower most layer is the outermost encapsulating scheme (see Figure 1.2). For example, in the case of TCP traffic over Ethernet, the data is first encapsulated with the TCP header (Layer 4), preceded by the IP header (Layer 3), which, in turn, is preceded by the Ethernet header (Layer 2).

Figure 1.2   Layering Encapsulation for a packet in the OSI Seven Layer Mode

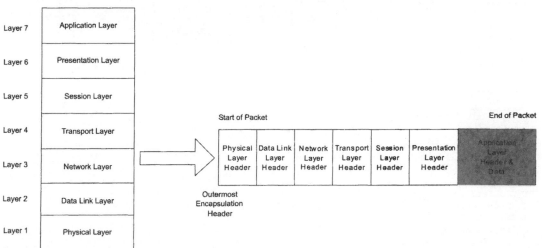

1.2 Communication Devices

Communication devices have a specific place in the network and implement specific protocols at each of the layers. A host system, for example, may implement all the layers of the OSI stack to communicate with a peer host entity. A hub or repeater may implement only the physical layer to regenerate the signals. An Ethernet switch may implement only physical and data link layer functionality. A router primarily operates at the network layer, so it will typically implement the first three layers, physical, data link, and network.

This section introduces a few of the popular communications devices and software that supports them. This will set the stage for subsequent chapters in the book. During these discussions, the TCP/IP protocol suite is assumed throughout the network. Figure 1.3 illustrates a typical network architecture, which is used as a reference for the following discussion on hosts, switches and routers. A host on LAN1 communicates with a server on LAN 2 across a WAN (Wide Area Network) using routers and switches. This topology and the individual devices used in it will form the basis for the discussions in later chapters. Specifically, the functionality and issues related to Layer 2 switches and routers (sometimes termed Layer 3 switches in this book) will form the underlying thread for the discussions on communications systems design and implementation.

Figure 1.3   A typical network architecture.

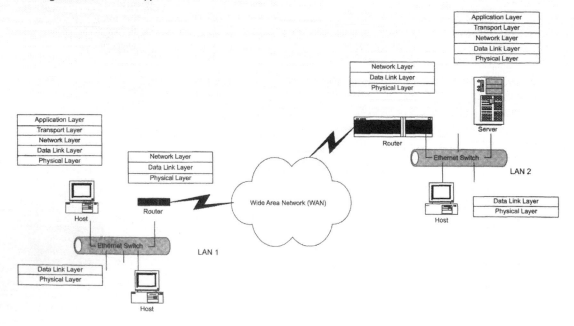

Host Systems

A host system connected to a LAN, such as Ethernet, communicates with devices on its own LAN and can also communicate with devices on other networks, such as Web servers. A host system with a Web browser uses an application protocol like HyperText Transfer Protocol (HTTP) to access the content provided by the Web server. In this situation, HTTP runs on top of TCP at the transport layer, which, in turn, runs over IP at the network layer. In Figure 1.3, the host software implements layers 1 through 5.

Layer 2 Switches

Layer 2 switches operate at the data link layer and switch MAC (Ethernet) frames between two LAN segments. They determine the destination Ethernet address from the MAC frame and forward the frame to the appropriate port. The port determination is done via a table which has entries in the form of a (Destination MAC Address, Port) pair. The Layer 2 switch constructs this table by learning the addresses of the nodes on each of its ports. It does this by monitoring traffic that the nodes send out on its LAN segment (see Figure 1.3). The source address on a MAC frame indicates that the node with this address is present on the LAN segment (and port) over which the frame was received.

Note that Layer 2 switches do not examine the network layer IP address of the packets they are forwarding, as these devices switch MAC frames between two hosts on the same IP network. Layer 2 switches are an evolution of the earlier transparent bridges. *Transparent* implies that the hosts are not aware of the existence of the switches—i.e., they do not target frames to the switches but only to the destinations on the same network. The switch makes decisions on forwarding the frame to the appropriate LAN segment without the source node being aware of its presence. For forwarding IP packets to other networks, a router is used.

Routers

Routers operate at the network layer of the OSI model and can forward IP packets between a source and destination. In the example of Figure 1.3, the host and the server with which it communicates are assumed to be on separate networks (also called subnets in the TCP/IP world). They are connected across a WAN (Wide Area Network) using routers.

Hosts send their "off-net" packets to the router, which, in turn, forwards these frames in the direction of the destination. If the destination host is directly connected, the packets are sent directly to the destination. If the destination host is not directly connected, the packets are sent to another router, from where they are directed towards the destination network, traversing, possibly, several intermediate networks. This is the scenario outlined in Figure 1.3.

IP routers build tables using routing updates that are exchanged between neighboring routers. These tables are used for forwarding IP packets across a network. The format and processing of routing updates are defined in routing protocol specifications like those for the Routing Information Protocol (RIP), Open Shortest Path First (OSPF), and Intermediate System–Intermediate System (IS-IS) protocols.

Using Multiple Protocols

Protocol functions constitute a large part of the software in a communications system. Software that implements protocol functionality is often called a *protocol stack*. Each protocol stack component performs the protocol function while stacked on top of or below another. For example, a TCP stack sits on top of an IP stack in a host implementation. The IP stack could sit on top of a PPP (Point to Point Protocol) stack for communicating over a serial interface.

Communications devices often perform more than one function—for example, a router may also perform Layer 2 switching. Also, there may be a need for the router to communicate as an end node for management (e.g., SNMP, Telnet, HTTP) purposes. In the case of an Ethernet Layer 2 switch and router, , the device needs to implement Layers 1 and 2 (Ethernet physical layer and Ethernet MAC layer) to perform the Layer 2 switching function. To realize the routing function, the device will implement Layers 1, 2, and 3 (Ethernet physical layer, Ethernet MAC layer, and IP layer). For the end node function, there is a need for Layers 1 through 4 (Ethernet physical layer, Ethernet MAC layer, IP layer, and TCP layer), as well as the application layer (realized in protocols like HTTP).

Telecom Equipment

Routers and switches are typically used in the data communications world, where data transfer is done via packet switching. In the telecommunications (telecom) world, circuit switching is the traditional switching scheme used for voice communications. In this scenario, the end node is typically a phone which is connected to the telecom network though a wired or wireless link. The connection from the phone terminates at a local exchange or switching center.

The phone links to the network by using messages sent via a tone (analog) or messages (digital). Multiple switching centers or exchanges use these messages to establish the connection from the dialing phone to the destination phone, a process known as *signaling*. The exchanges communicate with each other using a separate protocol called Signaling System #7 (SS7). This protocol enables the end-to-end connection between source and destination phones.

After the connection is established, voice conversation is carried as either analog or digital information. A supervisory function, known as *call control*, sets up and characterizes the connection between source and destination using a set of rules. For example, the destination phone may have specified that it is not willing to accept calls from the source phone. In this case, the exchange (switch) will deny the call, since it would have been informed of this rule through SS7.

In the scenario described, the handset or telephone is a low-complexity device, while the central office switch is at the other end of the spectrum. Commercial switches like the Lucent 5ESS™ and Nortel's DMS-100™ are extremely complex pieces of equipment which require hundreds of thousands of lines of software for call control and supervisory functions, which are the typical software components in the switch.

From an architectural perspective, circuit-switched systems can be designed such that the network contains the communications intelligence, and end systems have little or no intelligence. This is the typical view of the telecom world. The alternate is the view of the engineers in the IP/Internet world. In this view, end system devices may need to implement functions such as timeouts and retransmissions while the network performs basic forwarding functions, i.e., has little intelligence. Similar to the datagram-versus-virtual circuit argument, there is no clear winner.

Due to the increasing complexity of the software, the large circuit switching exchanges are being replaced by a new class of devices called "soft switches." These switches separate the control processing from data or payload processing. Complex control software can reside in an off-switch workstation, rather than an embedded device like the circuit switch. This strategy increases architectural flexibility, since the workstation software can be upgraded without affecting data processing. This separation of the control and data processing functions is common among recent communications software and systems.

Phones

Traditional phones use analog signals to communicate with the central office exchange. Some cordless phones still use analog methods to communicate with the central office to record, store, and retrieve messages. Office phones usually communicate with a local exchange located within the office, which is often called a private branch exchange (PBX) and is usually a scaled-down version of the central office exchange. Often the private exchanges use a proprietary digital communications and require software functions on them built into the phones. The private exchange will, in turn, use a digital method of communication with the central office exchange using a line like a T1 or E1 line.

Cellular phones now use digital communication with the network. These phones have a set of protocols starting from the physical layer that they use to communicate with the base transmission station (BTS). The software implementing these protocol stacks includes functionality to send periodic signals to the base transmission station to indicate the location of the phone within the cellular network as well as set up connections through the cellular network to another phone. While the cellular network is quite complex, some of the complexity is on the handset itself.

Convergence

Many companies typically have two networks — a telephone network and a data network. The telephone network is a circuit-switched network both internally and externally, while email, file transfer and collaborative work is done using IP on a packet-switched network. To reduce complexity, there is a movement towards a single, integrated network, where even voice is carried over the network in IP packets. In a typical voice-over-IP implementation, a phone is connected to an Ethernet network instead of a private branch exchange. Voice is sampled and digitized into discrete packets and sent over the Ethernet network. Packets are forwarded to reach destinations, which could be other IP phones or analog phones. With analog phones, a gateway converts the packets back into analog information to communicate with the analog phone.

Simple phones of the analog world are being replaced by a complex Ethernet phone, while the relatively homogeneous circuit-switching network is being supplanted by a more complex topology, including gateways and soft switches. All of these involve a fair degree of software complexity and introduce new system requirements. On a network where voice and data traffic travel over the same links, voice traffic must be given a higher priority, since a voice packet that arrives late is meaningless to the listener. In summary, phones are increasing in complexity with a need for protocol stacks to be implemented on them.

1.3 Types of Software Components

Broadly, there are two types of software components in a communications system: *protocol software*, which implements a protocol specification, and *systems software* (including infrastructure software) which usually includes a real-time operating system (RTOS) and an infrastructure to manage the hardware.

1.3.1 Protocol Software

Protocol software implements the protocol as detailed in a protocol specification. These specifications are usually specified by a standards body such as theInternational Organization for Standardization (ISO), Institution of Electrical and Electronic Engineers (IEEE), or International Telecommunications Union (ITU-T). Example protocols include the Internet Protocol (IP) and IEEE 802.2 Logical Link Control (LLC).

While the protocols are standard, implementation on communications systems varies widely. Small form factor devices such as PDAs (Personal Digital Assistant) and cell phones may limit the size of the protocol software resident on the device. On the other hand, carrier-class systems like edge routers may require that the protocol software be implemented in a distributed fashion.

Using the OSI layering as an abstraction mechanism, the software architecture of a complex communications system can be partitioned into higher and lower layers. For some functions, the higher and lower layers may be other protocols. For example, a device driver may be a lower layer function, while an application may be a higher layer implementation (see Figure 1.4).

Protocol implementation is typically based on a state event machine (SEM), also known as a *state machine*. The state machine is the core of the protocol implementation, typically in the form of the State Event Table (SET), which holds a set of rules specifying the action to be performed based on events. For example, a message is transmitted if a timeout occurs in a specific state.

Some parts of the protocol function may be augmented via hardware. A system implementing MPLS (Multi Protocol Label Switching) switching, for example, will provide the MPLS control protocols like LDP (Label Distribution Protocol) via software and perform the actual MPLS switching in hardware. Consequently, the software needs to be designed such that the interface between the software implementation and hardware implementations are clearly identified.

Figure 1.4   Protocol Implementation and Interfaces

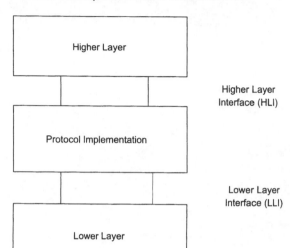

1.3.2 Infrastructure and Systems Software

The infrastructure and systems software components form the platform on which the communications system is built. Typically a real-time operating system (RTOS) forms the platform along with other software components such as buffer management, memory management, timer management subsystems, and device drivers. Protocols run on top of the operating system using the buffer management subsystem to exchange data on their interfaces and the timer management subsystem to keep track of timeouts.

In a large distributed architecture, communication systems can include multiple rack-mounted hardware modules. Infrastructure software handles the inter-board or inter-chassis communication and other software modules that monitor and manage the health of some hardware modules. For example, a shelf manager software component can run on one of the cards on the hardware shelf. The shelf manager may poll other cards on the shelf to ensure that they are functioning and that the communications link between the cards is operational. It may report status to an event manager, which forwards status to an external management entity.

The infrastructure components are typically the first pieces of software implemented during system development. These components need to be architected for performance and memory efficiency. In some cases, infrastructure components are available from the real-time operating system vendors in the form of specialized management libraries, while in other cases, they may need to be implemented by the engineers developing the software.

The Communications Ecosystem

The communications infrastructure requires a number of participants in the development and deployment process. The final product is sold by equipment vendors for use by end users and enterprises. The engineering effort in building the product requires tools at various levels. Figure 1.5 details the players in the food chain. They are:

- Electronic Design Automation (EDA) vendors
- Semiconductor component vendors
- RTOS, tools and software vendors
- Contract Manufacturers (CMs)
- Equipment Manufacturers (EMs)
- Home, Enterprise and service-provider users

Semiconductor component vendors supply their chips to the equipment manufacturers. Semiconductor vendors use tools from Electronic Design Automation (EDA) vendors to design chips, which are then supplied to the communications equipment manufacturers. If the EM staff designs some ASICs, these EDA vendors supply the tools to the equipment manufacturers also. The hardware designers on the EM staff design their hardware using these chips—they may include processors, network controllers, switch fabrics, network processors, and so on.

The processors will typically require an RTOS to run on them and may also include other software like third-party protocol stacks, redundancy framework software, and so on. These will be provided by the RTOS and third-party software vendors (or reused from earlier projects). The tool vendors provide the compilation, test and debugging tools for the engineering effort at the EM. The EMs design and develop the communications hardware and software using these tools. They can then sign up with a contract manufacturer (CM) to manufacture the actual equipment. These are then shipped to service-provider, enterprise, or home customers depending upon the distribution model. For example, a DSL service provider may take the complete responsibility and install the DSL modem at the customer premises. In this case, the EM never deals with the end customer.

Figure 1.5   Players in the communications infrastructure.

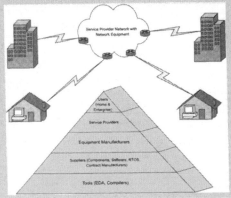

1.4 Design Considerations—A Prelude

- The following is an example of some of the issues faced by designers and developers of communications software. It is important that these issues are addressed early in the design for the product to meet its requirements.

- Does the hardware architecture involve one processor or multiple processors? Partitioning of functions will depend upon the number of processors.

- Has the RTOS been chosen? If so, does it provide memory protection between processes/tasks? The task partitioning and interfacing will depend upon this.

- What kind of performance is required? Can this be implemented completely in software, or does it require some hardware acceleration support?

- What are the protocols to be implemented, and how do they interface with the system software and other protocols? Can any of the protocols be accelerated via hardware?

- What kind of global data structures and tables are to be used for the implementation?

- What kinds of buffer and timer management are required?

- What kind of error notification and handling are required? Does the software require extensive configuration and control? What are the types of interfaces to be provided for such management?

- Does the RTOS provide a simulator which can be used for the development while the hardware is getting ready? What kind of testing can be done with the simulator? What kind of testing can be performed once the hardware is ready?

This book addresses these issues using the Layer 2/3 switch (or router) as an illustrative example. The aim is to provide the reader a flavor of the design considerations and tradeoffs in building communications software.

1.5 Summary

The OSI reference model is used for building communications systems and networks. Each device implements standard or proprietary protocols for the various layers that it implements. Each layer can be implemented via hardware and/or software. Hosts or end nodes originate and terminate communication, while network nodes like switches and routers transport the corresponding data.

The software implementation for a communications device involves two types of components: protocol software and infrastructure/systems software. Protocol software implements the software as detailed in the specification with higher layer and lower layer interfaces as well as state machines. System software includes the RTOS, drivers, buffer/timer management and other infrastructure functions.

1.6 For Further Study

Tanenbaum [2002] provides a significant amount of detail about the OSI model and the various protocols used in the networking world. Comer [2003] discusses network systems engineering with software and with network processors.

1.7 Exercises

1. How do you decide between implementing a protocol layer in software versus hardware? Enumerate some reasons.

2. Is power consumption a consideration in high-end systems? Or is it just confined to handhelds, PDAs, and cell phones? How can software be modified to address power consumption issues?

3. Discuss the advantages and drawbacks of an intelligent network (telecom view) and intelligent end nodes (IP view).

4. If you were to build a communications system, what are the pieces of infrastructure software that you would work on first? Give reasons.

CHAPTER 2

Software Considerations in Communications Systems

The previous chapter provided an introduction to communications systems and some of the issues related to the software on those systems. This chapter investigates software in greater detail, starting with an introduction to host-based communications along with a popular framework for building host-based communications software—the STREAMS environment.

Subsequently, we focus on embedded communications software detailing the real-time operating system (RTOS) and its components, device drivers and memory-related issues. The chapter also discusses the issues related to software partitioning, especially in the context of hardware acceleration via ASICs and network processors. The chapter concludes with a description of the classical planar networking model.

2.1 Host-Based Communications

Hosts or end systems typically act as the source and destination for communications. When a user tries to access a Web site from a browser, the host computer uses the Hyper Text Transfer Protocol (HTTP) to communicate to a Web server—typically another host system. The host system has a protocol stack comprising several of the layers in the OSI model. In the web browser, HTTP is the application layer protocol communicating over TCP/IP to the destination server, as shown in Figure 2.1.

Figure 2.1   Web browser and TCP/IP Implementation in Unix

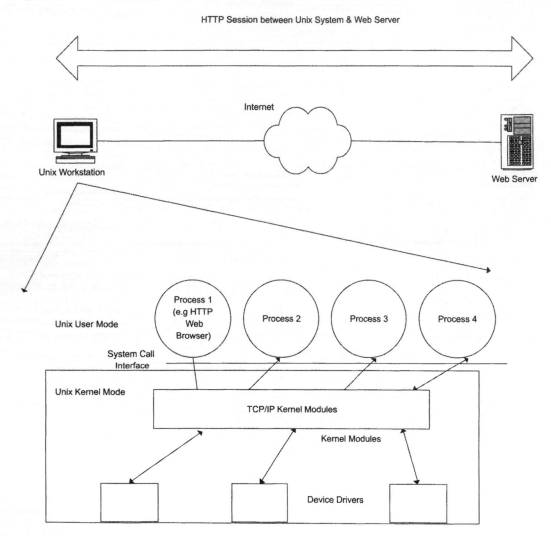

2.1.1   User and Kernel Modes

Consider a UNIX™-based host which provides the functionality of an end node in an IP network. In UNIX, there are two process modes—*user mode* and *kernel mode*. User applications run in user mode, where they can be preempted and scheduled in or out. The operating system runs in kernel mode, where it can provide access to hardware. In user mode, memory protection is enforced, so it is impossible for a user process to crash or corrupt the system. In this mode, processes cannot access the hardware directly but only through system calls, which, in turn, can access hardware. Kernel mode is, effectively, a "super user" mode. System calls execute in kernel mode, while the user process is suspended waiting for the call to return.

For data to be passed from user to kernel mode, it is first copied from user-mode address space into kernel-mode address space. When the system call completes and data is to be passed back to the user process, it is copied from kernel space to the user space. In the UNIX environment, the Web browser is an application process running in user mode, while the complete TCP/IP stack runs in kernel mode. The browser uses the `socket` and `send` calls to copy data into the TCP protocol stack's context (in the kernel), where it is encapsulated into TCP packets and pushed onto the IP stack.

There is no hard-and-fast requirement for the TCP/IP stack to be in the kernel except for performance gains and interface simplicity. In the kernel, the entire stack can be run as a single thread with its own performance efficiencies. Since the TCP/IP stack interfaces to multiple applications, which could themselves be individual user-mode processes, implementing the stack in the kernel results in crisp interfaces for applications. However, since memory protection is enforced between user and kernel modes, the overall system performance is reduced due to the overhead of the additional copy.

Some host operating systems like DOS and some embedded real-time operating systems (RTOSes) do not follow the memory protection model for processes. While this results in faster data interchange, it is not necessarily a reliable approach. User processes could overwrite operating system data or code and cause the system to crash. This is similar to adding an application layer function into the kernel in UNIX. Due to the system-wide ramifications, such functions need to be simple and well tested.

2.1.2   Network Interfaces on a Host

Host system network interfaces are usually realized via an add-on card or an onboard controller. For example, a host could have an Ethernet network interface card (NIC) installed in a PCI (Peripheral Component Interface) slot as its Ethernet interface. The Ethernet controller chip on the NIC performs the Media Access Control (MAC), transmission, and reception of Ethernet frames. It also transfers the frames from/to memory across the PCI bus using Direct Memory Access (DMA).

In the reception scenario, the controller is programmed with the start address in memory of the first frame to be transferred along with the default size of the frame. For standard Ethernet, this is 1518 bytes, including headers and Cyclic Redundancy Check (CRC). The controller performs the DMA of the received Ethernet frame into this area of memory. Once the frame transfer is complete, the actual size of the frame (less than or equal to 1518 bytes) is written into a header on top of the frame. At this stage, the Ethernet driver software needs to copy the data from the DMA area to be memory accessible to both the driver and higher layers. This ensures that subsequent frames from the controller do not overwrite previously received frames.

In the UNIX example, the Ethernet driver interfaces to the higher layer through a multiplexing/demultiplexing module. This module reads the protocol type field (bytes 13 and 14) in the Ethernet frame header and hands it off to the appropriate higher layer protocol module in the kernel. In a host implementing both IP and IPX, IP packets (protocol type 0x0800) are handed off to the IP module, while IPX packets (protocol type 0x8137) are handed off to the IPX module in the kernel (see Figure 2.2). The packets are subsequently processed by these modules and handed to applications in the user space. Frame transmission is implemented in a similar manner.

Figure 2.2   Unix Host implementing IP and IPX

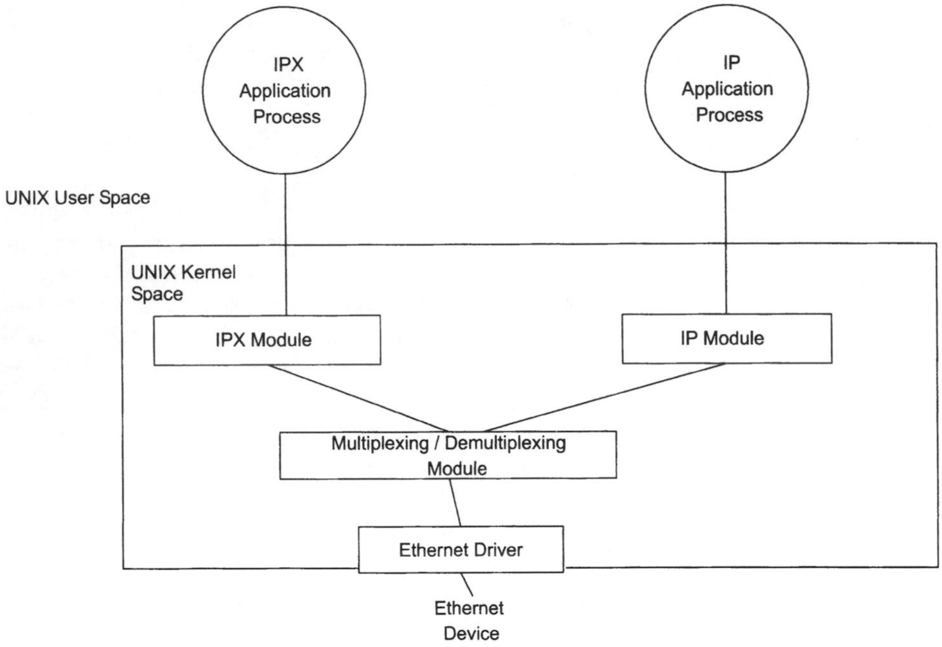

2.1.3   STREAMS Architecture

Using a modular approach permits the development of individual modules without the dependency on other modules. In the UNIX example, the IP module in the kernel may be developed independent of the Ethernet driver module, provided the interfaces between the two modules are clearly defined. When the two modules are ready, they can be combined and linked in. While this is a good mapping of the OSI model, an additional level of flexibility can be provided via an "add and drop" of functional modules while the system is running. The most common example of this is the STREAMS programming model, which is used in several UNIX hosts for implementing network protocol stacks within the kernel.

The STREAMS programming model was first specified in AT&T UNIX in the 1980s. It is now used in several UNIX systems and in some real-time operating systems (RTOSes) as well. STREAMS uses a model similar to the OSI layering model and provides the ability to create and manipulate *STREAMS modules*, which are typically protocol processing blocks with clear interfaces.

The model consists of a set of kernel-resident system calls, kernel resources, and utility routines to facilitate duplex processing and data paths between kernel mode and a user process. The fundamental unit of STREAMS is a stream, which is used for data transfer between a user process and a module in kernel space and can be decomposed into three parts: a Stream head, zero or more modules, and a driver.

Figure 2.3 shows the basic composition of a stream. The stream head is closest to the user process; modules have fixed functionality and interface to the adjacent layer (the stream head or another module). A module can be dynamically pushed on to or popped from the stream by user action, making this architecture suitable for the layering associated with communications protocols.

Figure 2.3   Stream components.

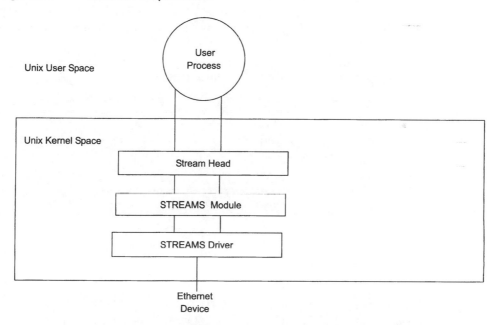

Each module has a "read" side and a "write" side. Messages traveling from the write side of a module to the write side of the adjacent module are said to travel downstream. Similarly, messages traveling on the read side are said to be traveling upstream. A queue is defined on the read and write side for holding messages in the module. This is usually a First In First Out (FIFO) queue, but priority values can be assigned to messages, thus permitting priority-based queuing. Library routines are provided for such operations as buffer management, scheduling, or asynchronous operation of STREAMS and user processes for efficiency.

STREAMS Messages

A message queue consists of multiple messages. Each message consists of one or more message blocks. Each message block points to a data block. The structure of a message block and a data block is shown in Listing 2.1. Most communications system suse this two-level scheme to describe the data transfer units. Messages are composed of a message block (msgb) and a data block (datab). The db_ref field permits multiple message blocks to point to the same data block as in the case of multicast, where the same message may be sent out on multiple ports. Instead of making multiple copies of the message, multiple message blocks are allocated, each pointing to the same data block. So, if one of the message blocks is released, instead of releasing the data block associated with the message block, STREAMS decrements the reference count field in the data block. When this reaches zero, the data block is freed. This scheme is memory efficient and aids performance.

STREAMS is important because it was the first host-based support for protocol stacks and modules which could be dynamically loaded and unloaded during program execution. This is especially important in a system that requires a kernel rebuild for changes to the kernel.

Listing 2.1   Streams message and data block structures.

```
struct msgb      *b_next;     /* Ptr to next msg on queue */
struct msgb      *b_prev;     /* Ptr to prev msg on queue */
struct msgb      *b_cont;     /* Ptr to next message blk */
unsigned char    *b_rptr;     /* Ptr to first unread byte*/
unsigned char    *b_wptr;     /* Ptr to first byte to write*/
struct datab     *b_datap;    /* Ptr to data block */
unsigned char    b_band;      /* Message Priority  */
unsigned short   b_flag;      /* Flag used by stream head  */
```

The data block organization is as follows:

```
unsigned char    *db_base;    /* Ptr to first byte of buffer */
unsigned char    *db_lim;     /* Ptr to last byte (+1) of buffer*/
dbref_t          db_ref;      /*Reference count- i.e.# of ptrs*/
unsigned char    db_type;     /* message type */
```

STREAMS buffer management will be discussed in greater detail in Chapter 6.

2.1.4   Socket Interface

The most common interface for kernel- and user-mode communication is the socket interface, originally specified in 4.2 BSD UNIX. The socket API isolates the implementation of the protocol stacks from the higher layer applications. A socket is the end point of a connection established between two processes—irrespective of whether they are on the same machine or on different machines. The transport mechanism for communication between two sockets can be TCP over IPv4 or IPv6, UDP over IPv4 or IPv6, and so on. A "routing" socket is also useful, for such tasks as setting routing table entries in the kernel routing table.

The socket programming model has been used in other operating systems in some popular commercial real-time operating systems such as Wind River Systems VxWorks™. This permits applications that use the socket API to migrate easily between operating systems that offer this API.

2.1.5   Issues with Host-Based Networking Software

Host-based networking software is not usually high performance. This is due to various reasons—there is very little scope for hardware acceleration in standard workstations, the host operating system may be inherently limiting in terms of performance (for example, the user and kernel space copies in UNIX), or the software is inherently built only for functionality and not for performance (this is often the case, especially with code that has been inherited from a baseline not designed with performance in mind). Despite these, protocols like TCP/IP in the UNIX world have seen effective implementation in several hosts.

Designers address the performance issue by moving the performance-critical functions to the kernel while retaining the other functions in the user space. Real time performance can also be addressed by making changes to the scheduler of the host operating system, as some embedded LINUX vendors have done.

2.2 Embedded Communications Software

Host machines running general-purpose operating systems are not the best platforms for building communications devices. Even though some routers are built on top of UNIX and Windows NT, they have seen limited use in the Internet. These routers perform all processing in software and have to work within the constraints of the general-purpose operating system, for example, equal-priority, timeslice-based scheduling, which can result in packet processing delays. Moreover, often times, these general-purpose systems are used to run other application code at the same time as the networking application.

The solution is to use dedicated communication hardware or an "appliance" which could be a router, switch, terminal server, remote access server, and so on. For our discussion, the dedicated appliance is an embedded communications device, to differentiate it from the host, which participates in the communication.

Common characteristics of a communications appliance are as follows:

- It typically runs a real-time operating system
- It has limited memory and flash
- It either has limited disk space or is diskless
- It provides a terminal and/or Ethernet interface for control and configuration

- It frequently has hardware acceleration capability

2.2.1   Real-Time Operating System

The real-time operating system (RTOS) is the software platform on which communications functionality and applications are built. Real-time operating systems may be proprietary (home grown) or from a commercial real-time operating system vendor such as VxWorks™ from Wind River Systems, OSE™ from OSE Systems, and Nucleus™ from Mentor Graphics.

The embedded market is still struggling with homegrown platforms. Often, engineers from the desktop world are surprised that so many embedded development engineers still develop their own real-time operating systems. Some commercial real-time operating systems have too much functionality or too many lines of code to be used for an embedded project. Perhaps more significant, many commercial real-time operating systems require high license fees. The per-seat developer's license can be expensive, and royalty payments may be required for each product shipped with the RTOS running on it. Some engineering budgets are too tight to handle this burden. Last, some engineers feel they can do a better job than an off-the-shelf software component—since the component may or may not be optimal for the application, especially if the RTOS vendor does not provide source code. However, this is not always a simple task.

RTOSes for Communications Systems

Early communications architectures were quite simple. The entire functionality was handled by a big dispatch loop—when data arrived, it was classified and appropriate actions performed. The operations were sequential, and there was no interaction between the modules which performed the actions. In this case, an RTOS would have been overkill and the dispatch loop was more than sufficient. However, communications systems are rarely this simple. They require interaction between functional modules at various layers of the OSI Seven Layer Model. If, say, TCP and IP were implemented as separate tasks, they would need an inter-task communication mechanism between them, which could be in the form of queues, mailboxes, or shared memory. If the two tasks needed to access some common data such as a table, semaphores might be used to protect access to the table. Commercial real-time operating systems provide these mechanisms and functions, and there is no need to recreate them.

RTOS vendors typically support the complete development environment, including the processor used for the project, the "tool chain" like compilers, assemblers, and debuggers, and other development tools. RTOS vendors have also tuned their operating systems to utilize some of the features of the processor being used for the development—something that would take developers on a homegrown project a much longer time to accomplish and is usually beyond the scope of the development project.

It is strongly recommended that developers of communication systems use commercially available real-time operating systems and place engineering focus on communications development rather than infrastructure.

Typically, homegrown projects end up adding more systems software to handle common infrastructure functions that the RTOS provides out of the box. Furthermore, newly developed OS software tends to be more complicated, convoluted, and buggy, making it more expensive to develop and maintain than the commercial RTOS. The following table summarizes the issues to consider for using a commercial RTOS instead of developing your own for the specific project.

Issue	Standard RTOS	Proprietary RTOS
Performance for a specific application	Less Optimized	More optimized
Maintenance	Responsibility of RTOS vendor	Responsibility of developer
Portability to multiple processors	Provided by RTOS vendor via separate packages	Has to be provided by developer for each processor
Support for standard Ethernet/serial devices	Provided as part of RTOS package for board	Has to be developed
Modifiability	Can be done only if RTOS vendor provides source code	Easily modifiable since source code is available internally
Tool Chain Support	Supported by RTOS vendor	Need to build using third-party development tools
Standard Interfaces, IPC mechanisms and APIs	Provided by RTOS vendor	Has to be designed in by the developer
Cost	High in dollar terms	Low in upfront dollar terms but could be high because of development effort/time and debugging

Board Support Package (BSP)

A typical embedded communications platform has a CPU, flash, or DRAM from which the code runs, DRAM for the data structures, and peripherals like serial and Ethernet controllers. The CPU requires its initialization sequence, after which it can perform a diagnostic check on the various peripherals. This software is CPU and board specific and is usually included in the RTOS specific to the board. This software, including the RTOS is also known as a Board Support Package (BSP). For Common Off The Shelf (COTS) communications boards, the BSPs are provided by the board manufacturer via a license from the RTOS vendor. Alternately, the RTOS vendor can provide the BSP for several popular COTS boards along with the RTOS developer license.

For boards developed internally, engineers have to create the BSP, which is not a trivial task. Most vendors offer board porting kits, which instruct engineers on how to create and test a homegrown BSP. RTOS vendors and commercial system integrators often provide board support as a consulting service.

Once the BSP is created, it allows an executable image of the RTOS or a portion of the RTOS to run on the target board and can be linked in with the communications application to form the complete image.

2.2.2 Memory Issues

Embedded communications devices rarely have a disk drive except when they need to store a large amount of data. These systems boot off a PROM (or flash) and continue their function. In a typical scenario, the boot code [see Figure 2.4] on flash decompresses the image to be executed and copies the image to RAM. Once decompression is complete, boot code transfers control to the entry point of the image in RAM, from which the device continues to function.

Figure 2.4   Boot sequence using ROM/Flash and RAM.

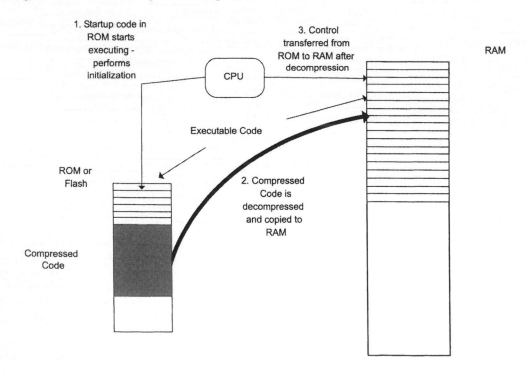

The embedded communications system comprises the RTOS and the communication application. The communication application is the complete set of protocol and systems software required for a specific function—for example, a residential gateway implementation. The software application and the RTOS use the RAM for data structures and dynamic buffer and memory requirements.

We earlier discussed memory protection in user and kernel modes in a UNIX host. The most significant difference between UNIX and the RTOSes is that there is no defined kernel mode for execution in RTOSes. Memory protection is enforced between tasks where appropriate, but that is all. Interestingly, many of the popular earlier RTOSes did not support memory protection between tasks. This did not limit the development engineers' flexibility during testing and debugging. However, in the field, these systems still ran the risk that user tasks with buggy code could corrupt the RTOS system code and data areas, causing the system to crash.

Recently, RTOSes with memory protection are seeing use in communications systems. Note that memory corruption bugs can be manifested in several indirect ways. For example, a task can overwrite data structures used to maintain the memory pool. The system will crash only when `malloc` is called next, causing much grief to the developer, who cannot figure out why the system has crashed and which task is responsible. Our recommendation is to use memory protection if it is available via a memory management unit (MMU) and in the RTOS.

Configuration and Image Download

A communications system needs to be able to save its current configuration so that if the system goes down, it can read the last saved configuration and use it to start up and be operational as soon as possible. For this reason, the communications device configuration is often stored in local non-volatile memory like an EEPROM or flash. Alternately, the configuration can be saved to and restored from (on restart) a remote host. In the same context, these systems also permit the download of a new image to upgrade the existing image on the flash. Most communications systems require an extensive amount of configuration to function effectively. For example, a router needs to be given the IP addresses of its interfaces; routing protocols need to be configured with timer values, peer information, and so on. Losing this configuration on reset would require the user to set these parameters all over again—which is not only a demand on network administrator time but also has a potential for misconfiguration.

Frequently, the flash system is used for field upgrades to avoid shipping the system back to the manufacturer for bug fixes and upgrades. To ensure that a new image does not cause the system to become non-operational, systems provide a backup to the last saved image. That is, there will be two sections on the flash, one for the existing saved image and the other for the newly downloaded image.

The new image is downloaded to a separate area of the flash so that the system can recover in case the download is unsuccessful or if the new image has problems. If the new image is downloaded and overwrites the existing image, the system cannot recover from this error and requires manual intervention. A key feature of telecom systems deployed in service provider networks, is their ability to perform a stepwise upgrade and rollback to the previous version of the software if the upgrade is unsuccessful.

2.2.3   Device Issues

Unlike the PCs or workstations, embedded communications devices do not come with a monitor and a keyboard. The only way to communicate with the "headless" embedded device is through a serial port or Ethernet. Headless devices can be booted and continue to operate without a keyboard and monitor. To communicate with the device through a serial port, you need to connect a terminal or use a PC with a terminal emulation program (like Hyperterminal on a Windows PC). The communications device typically has a Command Line Interface (CLI), which allows the user to type commands for configuring the device, displaying status and statistics, and so on.

In addition, an Ethernet port is typically used for communicating with the device for management purposes. The Ethernet port can be used to boot up a system as well as download new software versions. The TCP/IP stack runs over this Ethernet port, so that the network administrator can telnet to the device over this port to access the CLI. This facility is used by network administrators to manage the device from a remote location.

Device Drivers

Some RTOSes have their own communications stacks integrated or available as an additional package. These stacks interface to the hardware drivers through standard interfaces. Usually, the only hardware driver provided is the default Ethernet driver for the device used for management communication. Other drivers need to be written for specific ports to be supported by the application.

Standard driver interfaces ensure that the higher layer stacks such as IP will use the same interfaces irrespective of the device they are running on. For example, as in UNIX, some RTOSes use the following set of calls to access drivers, independent of the type of driver (Ethernet, Serial, and so on).

open ()   causes the device to be made active
close ()   results in the device being made inactive
read ()   is used for reading data received by the device
write ()   is used for writing data to the device
ioctl ()   is used for configuring and controlling the device

Applications using this interface will not need to be modified when moving to a new platform, as long as the new driver provides the same interface.

2.2.4   Hardware and Software Partitioning

Partitioning functionality between hardware and software is a matter of technology and engineering with the constraints of optimization. For example, a DSL modem/router may implement some compression algorithms in software instead of hardware, to keep the cost of hardware lower. The application environment may not consider this as a restriction, if the performance of such a software implementation is sufficient for normal operation. As technology progresses, it may be easier and less expensive to incorporate such functionality into a hardware chip. Developers should provide flexibility in their implementation via clearly defined interfaces, so that an underlying function can be implemented in software or hardware.

Size–Performance Tradeoff

Students of computer science and embedded systems would be familiar with issues related to size and performance and how they sometimes conflict with each other. Caching is another size–performance tradeoff made in the embedded communications space. A lookup table can be cached in high-speed memory for faster lookups, but instead of caching the entire table, only the most recently accessed entries could be cached. While this improves the performance of the system, there is an associated cost in terms of additional memory usage and the complexity of the caching algorithms.

Depending on application and system requirements, memory is used in different ways. For example, the configuration to boot up a system could be held in EEPROM. The boot code can be in a boot ROM or housed on flash. DRAM is typically used to house the executable image if the code is not executing from flash.

DRAM is also used to store the packets/buffers as they are received and transmitted. SRAM is typically used to store tables used for caching, since caching requires faster lookups. SRAM tends to be more expensive and occupies more space than DRAM for the same number of bits.

High-speed memory is used in environments where switching is performed using shared memory. Dual-port memory is used for receive and transmit buffer descriptors in controllers such as the Motorola PowerQUICC™ line of processors.

For cost-sensitive systems, the incremental memory cost may not be justifiable even if it results in higher performance. Similarly, when performance is key, the incremental complexity of a design that uses SRAM for caching may be justified.

Fast Path and Slow Path

When designing communications systems, the architecture needs to be optimized for the *fast path*. This is the path followed by most of (the normal) packets through the system. From a software perspective, it is the code path optimized for the most frequently encountered case(s).

Consider a Layer 2 switch which needs to switch Ethernet frames at Layer 2 between multiple LAN segments. The same switch also acts as an IP end node for management purposes. The code path should be optimized so that the switching is done at the fastest rate possible, since that is the main function of the system. If some of the Ethernet frames are destined to the switch itself (say, SNMP packets to manage the switch), these packets will not be sent through the fast path. Rather, they will be processed at a lower priority, i.e., they will follow the *slow path*.

Host Operating Systems versus RTOSes

Host operating systems like UNIX or LINUX are seeing deployment in some embedded communications devices, though they were not originally tuned for real-time applications. The following provides a checklist about the issues to consider when choosing between host and real-time operating systems. The Linux operating system is chosen as an example of a desktop OS for this purpose.

Evaluation Criterion	Choice of Linux or RTOS
Does the application "really" need real-time performance? E.g., if the application tasks are scheduled only periodically and most of the time-critical functions are handled via hardware, then there is really no need to go for an RTOS.	Linux
Offer standard APIs (like the socket API) for applications	Linux or embedded RTOS (if it offers the same APIs)
Possibility of modifying the Linux scheduler to be "closer" to real time	Linux, if the modification is possible
Availability for a specific hardware platform	Commercial RTOSes support more platforms
Tool Chain Support	Commercial RTOSes have better tool chain integration
Cost	Linux has no upfront fee or royalties

From the table, it is clear that an open source operating system like Linux is a growing threat to the RTOS business. We recommend that developers choose an OS platform by using evaluation criteria similar to the ones above.

Another example of fast- and slow-path processing is the handling of IP packets with options in a router. The router normally forwards IP packets from one interface to another based on the destination address in the packet. However, the IP protocol defines some optional parameters, called IP options, that can be present in the IP header. One such option is the *Record Route* option, where the router has to record its IP address in the designated space in the IP header. This will indicate that this router was on the path that the packet took to reach the destination. Options are typically used for diagnostic purposes; most packets will not include options.

If the IP forwarding logic is done in software, the fast-path software will be the one optimized to handle packets without options. If this software sees a packet with options, it will hand the packet off to the slow-path software and return, so that it can process the next packet. The slow-path software, typically in a separate task, will handle the packet when it is scheduled.

The separation between the fast path and slow path is the basis for hardware acceleration, discussed next.

2.2.5 Hardware Acceleration

All the information presented earlier about networking software assumed that it runs on a single, general-purpose processor (GPP). These are the processors like the ones used in workstations. They include the MIPS™ and PowerPC™ line of processors, which have a strong RISC (Reduced Instruction Set Computer) bias.

While these processors are powerful, there is always a limit to the performance gain from a software-only implementation. For devices with a small number of lower speed interfaces like 10/100 Mbps Ethernet, these processors may be sufficient. When data rates increase, and/or when there are a large number of interfaces to be supported, software based implementations are unable to keep up.

To increase performance, networking equipment often includes hardware acceleration support for specific functions. This acceleration typically happens on the fast-path function. Consider a Layer 2 switch, which requires acceleration of the MAC frame forwarding function. A Gigabit Ethernet switch with 24 ports, for example, can be built with the Ethernet switching silicon available from vendors like Broadcom, Intel or Marvell along with a GPP on the same board. The software, including the slow-path functions, runs on the GPP and sets up the various registers and parameters on the switching chipset.

These chips are readily available for any Ethernet switch design from the semiconductor manufacturer and are known as *merchant silicon*. These chips are available with their specifications, application notes, and even sample designs for equipment vendors to build into their system. In the case of an Ethernet switching chip, the device specifies connectivity to an Ethernet MAC or to an Ethernet PHY (if the MAC is already included on the chip, as is often the case). Hardware designers can quickly build their boards using this silicon device.

The Broadcom BCM5690 is one such chip. It implements hardware-based switching between 12 Gigabit Ethernet ports, so two 5690s can be used to build a 24-port Gigabit Ethernet switch. Once the 5690s have been programmed via the CPU, the switching of frames happens without CPU intervention.

Software vendors are provided with the programming interfaces to these devices. For example, for a switching device, we may be able to program new entries into a Layer 2 forwarding table. These interfaces may be provided through a software library from the silicon vendor when the vendor does not want to disclose the internal details of the chip. Often, the details of the chip are provided, so software engineers can directly program the registers on the chip for the required function. Independent of the method used to program the device, software performance can be enhanced by offloading the performance-intensive functions to the hardware device.

In summary, *hardware acceleration is used for the fast-path processing.*

ASICs

Not all hardware acceleration devices are available as merchant silicon. Some equipment vendors believe that merchant silicon does not address all their performance requirements or support the number of ports required. For example, the design may require that we need to support MAC and switching functionality for 48 Gigabit Ethernet ports on a single line card. Merchant silicon may not be able to satisfy this requirement, so an Application Specific Integrated Circuit (ASIC) needs to be developed. While designing this chip, engineers can add functionality specific to their system—in our example, this can include additional functionality for Layer 3 and 4 switching, which may not be available in merchant silicon.

While an ASIC is very efficient for the functions needed, it is quite expensive to develop. It typically takes about nine months to develop and, depending upon the tools and the engineering effort needed, can run into even millions of dollars. The upside is that the equipment vendor now has a proprietary chip which provides superior functionality/performance to any merchant silicon and thus provides a competitive differentiation. Several communications equipment vendors do not use merchant silicon for their core products. They maintain large engineering teams dedicated to working on custom chips. Also, when the technology is not proven, or the vendor has a proprietary twist on a technology, ASICs are commonly used.

Some merchant silicon, like the Broadcom BCM5690 and Marvell Prestera line of products, are also known as *Net ASICs* or *configurable processors*, to distinguish them from network processors, discussed next.

Network Processors

Network processors (NPs) are another type of network acceleration hardware, available from vendors such as Agere, AMCC, IBM, Intel, and Motorola. A network processor is simply a *"programmable ASIC,"* which is optimized to perform networking functions. At high speeds, a processor has very little time to process a packet before the next packet arrives. So, the common functions that a packet processor performs are optimized and implemented in a reduced instruction set in a network processor. The performance of an NP is close to that of an ASIC; it has greater flexibility since new "microcode" can be downloaded to the NP to perform specific functions.

Programmable hardware is important because networking protocols evolve requiring changes in the packet processing. The hardware needs to analyze multiple fields in the packet and take action based on the fields. For example, for an application like Network Address Translation (NAT), there may be fields in the packet which need to manipulated to reflect an address change performed by the address translation device. These are implemented via functions called Application Layer Gateways (ALGs) which are present in the address translation device implementing NAT. ALGs are dynamic entities and more are defined as applications are developed. Since the address translation performance should not be degraded as more and more ALGs are added, programmable hardware like network processors are a good fit for implementing NAT.

2.2.6 Control and Data Planes

A high level method of partitioning the functionality of the system is by separating it into functions that perform:

1. All the work required for the basic operation of the system (e.g. switching, routing)
2. All the work required for (1) to happen correctly

The classical planar networking architecture model uses this partitioning scheme to separating the communications functionality into three distinct planes (see Figure 2.5):

- Control Plane
- Data Plane
- Management Plane

Figure 2.5   Classical planar networking architecture.

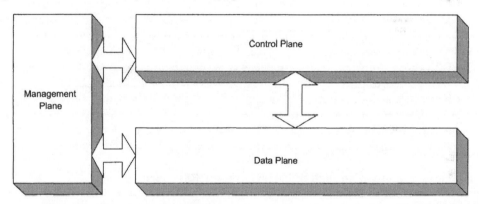

The data plane is where the work required for the basic operation takes place. The control and management planes ensure that the data plane operation is accurate.

The control plane is responsible for communicating with peers and building up tables for the data plane to operate correctly. The functions include peer updates, signaling messages, and algorithmic calculations. The control plane functions are typically complex and might also involve conditional execution. Examples of control plane protocols include the Open Shortest Path First (OSPF), Signaling System 7 (SS7) in the telecom world, signaling protocols like Q.933 in Frame Relay, and so on.

The data plane is responsible for the core functions of the system. For example, in the case of a router, the data plane performs the IPv4 forwarding based on tables built up by control protocols such as OSPF. Data plane functions are relatively simple and repetitive and do not involve complex calculations. There is little or no conditional execution required on the data plane. The functions on the data plane are relatively simple and do not involve complex calculations, as in the control plane (e.g., the OSPF protocol requires a complex Shortest Path First calculation based on information obtained from protocol updates).

The management plane spans across both the control and data planes and is responsible for the control and configuration of the system. This is the part of the system which performs housekeeping functions. It also includes functions to change configuration and to obtain status and statistics. Functions like SNMP, Command Line Interface (CLI), and HTTP-based management operate on the management plane.

2.2.7 Engineering Software for Hardware Acceleration

Hardware acceleration is typically used in the data plane and typically for fast-path processing. Software on the data plane is responsible for initializing the hardware used for acceleration, configuration, and programming. It also handles the slow-path processing.

While writing communications software for the data plane, it is essential that we partition functionality so that hardware acceleration can be added very quickly. Consider an IPSec implementation. IPSec is used in the TCP/IP world for securing communications between hosts or between routers. It does this via adding authentication and encryption functionality into the IP packet. The contents of the authentication and encryption headers are determined by the use of security algorithms which could be implemented in hardware or software. An algorithm like Advanced Encryption Standard (AES) can be implemented in a security chip, like those from Broadcom and HiFn.

An IPSec implementation should be written only to make function calls to the encryption or authentication algorithm without the need to know whether this function is implemented in software or hardware. In Figure 2.6, an encryption abstraction layer provides this isolation for the IPSec module; the IPSec module will not change when moving to a new chipset or software-based encryption.

Figure 2.6   Encryption abstraction layer for an IPSec module.

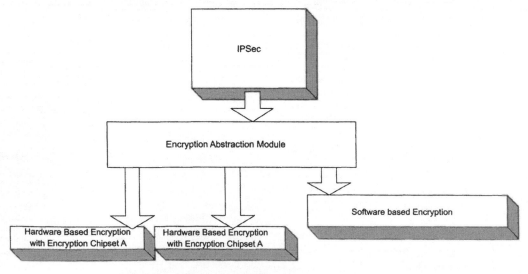

Software with Hardware Acceleration—A Checklist

The following is a checklist for engineering software in the presence of hardware acceleration. The underlying premise is that the software is modular, so it will be efficient with and without acceleration.

- Design the code to be modular so that functions in the data plane can be easily plugged in or out depending upon hardware or software implementation.
- Separate the fast-path and slow-path implementation of the data plane up front
- Maximize performance in the data plane even if it is a software-only implementation—it is very likely that the software implementation is sufficient for some low-end systems. This includes efficient search algorithms and data structures.
- Handle all exception processing in software
- Ensure that interrupt processing code is efficient—for example, read counters and other parameters from the hardware via software as fast as possible, since they run the risk of being overwritten.
- Do not restrict the design such that only certain parts of the data plane may be built in hardware. Network processor devices can move several data plane functions into software.
- Ensure that performance calculations are made for the system both with and without hardware acceleration—this is also one way to determine the effectiveness of the partitioning.
- When interfacing to hardware, use generic APIs instead of direct calls to manipulate registers on the controller. This will ensure that only the API implementations need to be changed when a new hardware controller is used. Applications will see no change since the same API is used.

The use of hardware acceleration is a function of the product requirements and the engineering budget. While it is certainly a desirable, developers should not depend upon it as a way to fix all the performance bottlenecks in their code.

2.3 Summary

Hosts, which communicate with peers, implement their networking software functions over a host operating system like UNIX. The functions can be in the user mode or kernel mode. STREAMS is an example of a framework for building networking stacks in the UNIX environment. The performance of host-based networking software is not high, due to several reasons—restrictions of the operating system, legacy code, kernel- to user-mode transitions, and so on.

Embedded communications devices typically use a real-time operating system as the platform for building their software functions. The RTOSes may be available from vendors or built in house. Using a commercial RTOS with standard APIs permits application portability. Memory considerations like booting, flash, and DRAM versus SRAM are all important in embedded communications systems.

An implementation of the system will require clear partitioning of the control plane, data plane, and management plane. In the data plane, both fast-path and slow-path processing will need to be identified. Hardware acceleration via ASICs and NPs can be used for specific functions in the data plane fast path.

2.4 For Further Study

Berger [2002] discusses the hardware–software partitioning issue as well as checklist for choosing an RTOS. Comer [2003] details the need for custom silicon like ASICs and network processors.

2.5 Exercises

1. List reasons why you think a commercial communications appliance can use a host operating system as the base OS.
2. Determine which RTOS vendors support a STREAMS framework either directly or from third-party vendors.
3. How does a security chip work in conjunction with a network processor? Provide a simple diagram of the solution.
4. Detail the advantages and disadvantages of using a configurable ASIC versus a programmable network processor.
5. A powerful hardware acceleration device is being considered by the hardware team. The problem is that its APIs and documentation are limited. The hardware and system engineers are going ahead with it despite these limitations. Explain how you would approach this situation and the steps you would take to de-risk the software development using this chip.

CHAPTER 3

Software Partitioning

This chapter addresses issues related to partitioning in communications software, both in terms of functionality and system implementation. While the OSI model can be used to partition the protocol software functionality, it is very rigid in terms of design. For performance reasons, some visibility into the higher and lower layers may be required, as explained below.

The discussion then moves on to the delineation between tasks and modules and how a typical communications system is implemented. An example of a Layer 2 switch is used to outline the various modules—including the device driver, protocol tasks and system/management modules. The chapter concludes with a description of the inter-module interfaces, including the use of standards-based and proprietary interfaces.

3.1 Limitations of Strict Layering

Chapter 1 detailed the OSI Seven Layer Model. The simplest way to partition the system is to follow this model by designing each protocol layer as a single module with clear interfaces with its upper and lower layers. If strict layering is followed, each protocol module will be unaware and independent of its upper and lower layers. This, however, is difficult to implement in practice, the key reasons being:

- Protocol Dependencies
- Performance Considerations
- Hardware and System Configuration

Protocol Dependencies

Consider the case of a host implementing the TCP/IP stack running over Ethernet. The TCP stack sits on top of the IP stack, which, in turn, sits above the Ethernet driver. The frame formats are shown in Figure 3.1. The destination and source addresses are 6 bytes each and occupy the first 12 bytes of the frame. The two-byte "type" field occupies bytes 13 and 14. This field is followed by the IP header, the TCP header, and data.

Figure 3.1   TCP/IP packets.

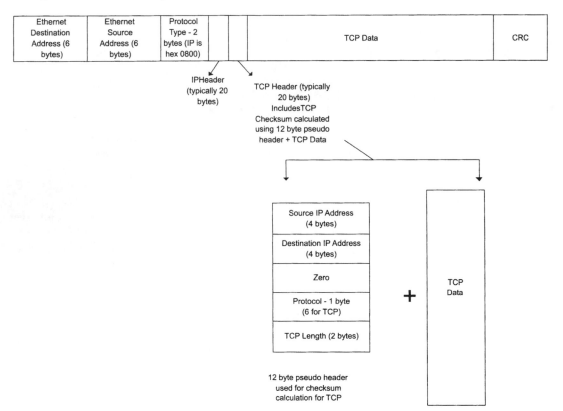

The Ethernet driver does not require any knowledge of the IP layer or headers, except that it needs to know that the protocol is IP, so that it can insert the correct frame type in the two-byte Ethernet type field. This can be implemented with a function call to the driver which passes the frame type and a pointer to the IP packet. The Ethernet hardware calculates the checksum for the complete Ethernet frame, including the header and transmits the entire packet over the Ethernet line via the Ethernet Media Access Control (MAC).

However, on the TCP side, the issue is more complex. Before a TCP segment (the data component of the TCP packet) can be sent, it calculates the checksum for the data in the segment. Prior to this calculation, the TCP function adds a "pseudo header" on top of the data. This pseudo header includes the source IP address and the destination IP address, along with the protocol type (number 6, which is the IP protocol type for TCP). The pseudo-header format and the TCP header are shown in Figure 3.1.

The source and destination IP addresses required by TCP for the pseudo header are the responsibility of he IP layer. With strict layering, the IP address information should not be visible at the TCP layer. However, since the TCP checksum requires these addresses, the TCP layer has to obtain the information from the IP layer—a violation of strict layering.

Performance Considerations

Consider the case of a TCP packet to be transmitted out on an Ethernet interface. The IP header is added before the TCP header, after which the Ethernet header is added before the IP header. Note that the data is provided by the TCP module and the remaining modules only add their headers before the frame. We can avoid copying the frame by creating the TCP packet such that it can accommodate the IP header and the Ethernet header. This is done by starting the TCP packet at an offset, which is calculated by adding the Ethernet header and the IP header sizes to the start of the buffer. This requires the TCP layer to have knowledge of the IP header and Ethernet header sizes—which again deviates from strict layering principles.

Hardware and System Configuration

Layering is also subject to the hardware and system configuration. For example, we may split TCP functionality to provide termination and data transfer on a line card while performing the control processing on a separate control card. Similarly, the encapsulation scheme may put a new twist on layering—there are cases where a frame relay packet may be encapsulated inside PPP (Point to Point Protocol—used on serial links to carry Layer 3 and Layer 2 traffic) and other cases where a PPP packet may be encapsulated inside a frame relay packet.

The conclusion is that while layering is a good way to partition the protocol functionality, we may not always be able to implement it in a strict fashion.

3.2 Tasks and Modules

The simplest way to partition the software is to decompose it into functional units called *modules*. Each module performs a specific function. An Ethernet driver module is responsible for configuration, reception, and transmission of Ethernet frames over one or more Ethernet interfaces. A TCP module is an implementation of the TCP protocol.

A module can be implemented as one or more tasks. We make the distinction between a task and a module as:

A module is a unit implementing a specific function. A task is a thread of execution.

A thread is a schedulable entity with its own context (stack, program counter, registers). A module can be implemented as one or more tasks (i.e., multiple threads of execution), or a task can be implemented with multiple modules. Consider the implementation of the IP networking layer as an IP task, an ICMP (Internet Control Message Protocol) task, and an ARP (Address Resolution Protocol) task. Alternately, the complete TCP/IP networking function can be implemented as a single task with multiple modules—a TCP module, an IP module, and so on. This is usually done in small–form factor devices, where both memory requirements and context switching overheads are to be minimized.

3.2.1 Processes versus Tasks

In desktop systems, a process represents the basic unit of execution. Though the terms *process* and *task* are used interchangeably, the term "process" usually implies a thread of execution with its own memory protection and priority [see Figure 3.2]. In embedded systems, the term "task" is encountered more often. Tasks do not have memory protection—so a task can access routines which "belong" to another task. Two tasks can also access the same global memory; this is not possible with processes.

Figure 3.2   Processes and tasks.

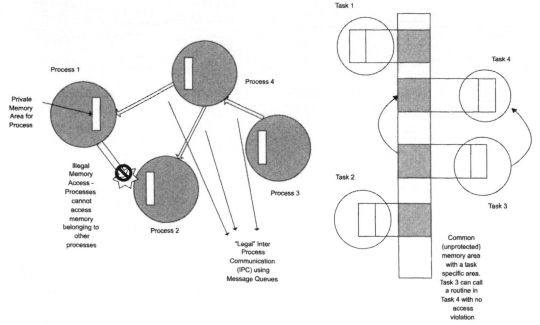

A. Processes with memory protection   B. Tasks with no memory protection

A process can also have multiple threads of execution. A thread is a *lightweight process*, which can share access to global data with other threads. There can be multiple threads inside a process, each able to access the global data in the process. Thread scheduling depends upon the system. Threads can be scheduled whenever the process is scheduled with round-robin or thread priority scheduling, or they can be part of a system-wide thread scheduling pool. A key point to note is:

In embedded systems with no memory protection, a task is the equivalent of a thread—so we can consider the software for such an embedded system as one large process with multiple threads.

3.2.2 Task Implementation

Independent functions and those that need to be scheduled at different times can be implemented as tasks, subject to the performance issues related to context switching. Consider a system with eight interfaces, where four of the interfaces run Frame Relay, while the other four run PPP. The two protocols are functionally independent and can run as independent tasks.

Even when tasks are dependent upon each other, they may have different timing constraints. In a TCP/IP implementation, IP may require its own timers for reassembly, while TCP will have its own set of connection timers. The need to keep the timers independent and flexible, can result in TCP and IP being separate tasks.

In summary, once the software has been split into functional modules, the following questions are to be considered for implementation as tasks or modules:

1. Are the software modules independent, with little interaction? If yes, use tasks.
2. Is the context-switching overhead between tasks acceptable for normal operation of the system? If not, use modules as much as possible.
3. Do the tasks require their own timers for proper operation? If so, consider implementing them as separate tasks subject to 1 and 2.

3.2.3 Task Scheduling

There are two types of scheduling that are common in embedded communications systems. The first and most common scheduling method in real-time operating systems is *preemptive priority-based scheduling*, where a lower priority task is preempted by a higher priority task when it becomes ready to run. The second type is non-preemptive scheduling, where a task continues to run until it decides to relinquish control. One mechanism by which it makes this decision is based on CPU usage, where it checks to see if its running time exceeds a threshold. If that is the case, the process relinquishes control back to the task scheduler. While non-preemptive scheduling is less common, some developers prefer preemptive scheduling because of the control it provides. However, for the same reason, this mechanism is not for novice programmers.

Several commercially available RTOSes use preemptive priority-based scheduling with a time slice for tasks of equal priority. Consider 4 tasks, A, B, C, and D, which need to be scheduled. If Task A, B, and C are at a higher priority than Task D, any of them can preempt D. With the time slice option, A, B, and C can all preempt Task D, but if the three tasks are of equal priority, they will share the CPU. A time slice value determines the time before one equal-priority task is scheduled out for the next equal-priority task to run.

For our discussions, we will assume hereafter the preemptive priority-based scheduling model. Time slicing is not assumed unless explicitly specified.

3.3 Module and Task Decomposition

This section will discuss design considerations with respect to partitioning the system into tasks and modules. It will then use the example of a Layer 2 Ethernet switch to illustrate the concepts. The assumption is that the system is a single-processor system running an RTOS. Systems with multiple boards and processors are discussed in Chapter 8.

The following list is a set of guidelines for organizing the modules and tasks in a communications system:

1. There are one or more drivers which need to handle the various physical ports on the system. Each driver can be implemented as a task or module. There are one or more Interrupt Service Routines (ISRs) for each driver.

2. The drivers interface with a higher layer demultiplexing/multiplexing task for reception and transmission of data. This is the single point of entry into the system for received frames. Similarly, this task is the only entity which interfaces to the driver for transmission.

3. Each protocol is designed as a module. It can be implemented as a task if it requires independent scheduling and handling of events, such as timers. If the overhead of context switching and message passing between the tasks is unacceptable, multiple protocol modules can run as one task.

4. Control and data plane functions, along with fast/slow path considerations will be used for grouping tasks and modules. Hardware acceleration will also play a part here.

5. Housekeeping and management plane functions like SNMP agents and Command Line Interface (CLI) will need to have their own task(s).

6. If memory protection is used, the interfaces between the tasks will be via messages. Otherwise, functional interfaces can be used. This will be discussed in greater detail in Section 3.6

The above guidelines are applied in the example of the Layer 2 switch, as discussed below.

3.4 Partitioning Case Study—Layer 2 Switch

Consider a Layer 2 Ethernet switch, which switches Ethernet frames between ports. This example system consists of 8 Ethernet ports for switching traffic and a management Ethernet port. It contains a CPU which runs the software for the system. For now, assume that there is no hardware acceleration and that all the switching is done by software running on the CPU.

Figure 3.3 shows the software architecture of the device. It requires the following:

- Driver(s) to transmit and receive Ethernet frames from the ports
- A set of modules/tasks to run the protocols required for the switch
- A set of modules/tasks required for the system operation and management

Figure 3.3   Typical Architecture of a Layer 2 Switch.

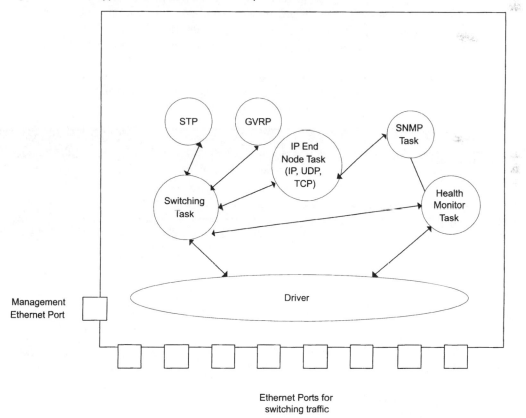

Ethernet Ports for
switching traffic

3.4.1 Device Driver

The device driver is the module closest to the hardware and is responsible for transmission and reception of frames. For transmission, the driver obtains the data from its higher layer, while, for reception, the driver needs to pass the received frame to its higher layer. The driver typically handles frames received from the hardware controller by either polling or by interrupts generated when a frame is received. The method used depends on application requirements.

With polling, the driver polls the device for received frames at periodic intervals. If the device indicates that frames have been received, the driver reads these frames from the device, frees buffers for the device to read frames into, and passes the frames to the higher layer. With interrupts, the driver receives an indication that a frame has been received, based on which it can read the frames from the device.

Polling is a relatively inefficient way to handle frames, especially if frames arrive infrequently. The driver polls the device even when no frames have been received, expending processing time which could be spent in other operations. Interrupts are more efficient but can easily swamp the system when frames arrive at a rapid rate. In the Layer 2 switch, frames can arrive at a maximum rate of 14,880 64 byte frames/second on a 10 Mbps Ethernet. If an interrupt is generated for each of the frames, it can easily overload the system. To handle this, drivers use a combination of interrupts and polling in conjunction with the controller, as described below.

Optimization of Reception

Several techniques are used to optimize controllers for receiving frames. Consider a case in which a list of multicast Ethernet addresses are to be recognized. In the Layer 2 switch, one such multicast address is used for Spanning Tree Protocol (STP) frames. STP is used to ensure that loops are detected and eliminated in the bridge topology. It involves periodic transmission of frames between bridges.

The STP frames use the multicast MAC address 01-80-C2-00-00-00, so the Ethernet controller needs to receive and pass up all MAC frames with this multicast address in the destination address field, whenever STP is enabled. The controller can perform this match on chip registers which store the MAC addresses. The driver programs this register with the MAC addresses based on the higher layer configuration. The higher layer configuration can take place via a Command Line Interface (CLI) with a command like "Enable STP", which, in turn, causes the "enable" event to be propagated all the way to the driver to receive STP multicast frames. Thus, higher layers need not be aware of the details at the driver level.

Frame Reception

In Figure 3.4, a driver provides receive buffers to the controller. These buffers are located in system memory, and the controller can access these receive buffers to directly DMA the received Ethernet frames. Typically, the controller manages the buffers in such a way that, if the full frame does not fit into the first buffer, it reads a portion of the frame into the first and copies the remaining into the second buffer, and so on. The amount of "overflow" into the next buffer will be determined by the size of the frame and the size of the buffer.

Figure 3.4   Frame Reception and Buffer Handling

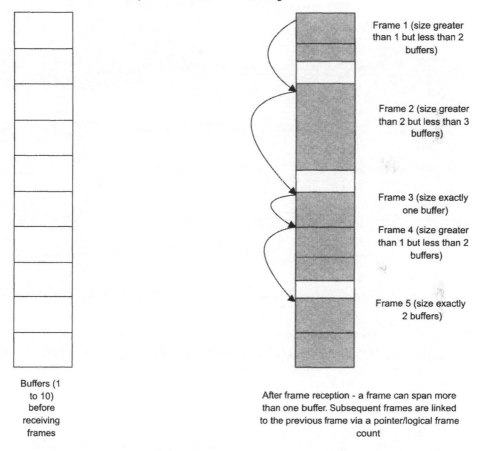

Frame 1 (size greater than 1 but less than 2 buffers)

Frame 2 (size greater than 2 but less than 3 buffers)

Frame 3 (size exactly one buffer)

Frame 4 (size greater than 1 but less than 2 buffers)

Frame 5 (size exactly 2 buffers)

Buffers (1 to 10) before receiving frames

After frame reception - a frame can span more than one buffer. Subsequent frames are linked to the previous frame via a pointer/logical frame count

In Figure 3.4, five frames of different sizes are received into the buffers. The figure shows that a frame "points" to the buffer housing the start of the next frame. This can be done using a pointer or an end-of-frame indication. Using Figure 3.4, assume a buffer size of 258 bytes, of which two bytes are used for management or "housekeeping" purposes, and 256 bytes are used for data. The parameters can include a count of the valid bytes in the buffer, whether this is the last buffer in the frame, status of the reception, and so on.

Now assume that the frames received are of size 300, 600, 256, 325 and 512 bytes. Using modulo arithmetic (i.e., if there is a remainder when dividing the frame size by the buffer size, you need to add 1 to the number of buffers needed), we need 2, 3, 1, 2 and 2 buffers respectively for the five received frames (see Table 3.1). The "Total Frame Size" column details how the frame sizes are calculated based on the count of valid bytes in the buffers constituting the frame (the last buffer in a frame is the one where the end-of-frame indication is set by the controller).

Table 3.1   Frame buffers.

Buffer No.	Count of Valid Bytes	Frame Reference	Total Frame Size
1	256	Frame 1	
2	44	Frame 1 and end of frame	Buffer 1 count + Buffer 2 count = 300 bytes
3	256	Frame 2	
4	256	Frame 2	
5	88	Frame 2 and end of frame	Buffer 3 count + Buffer 4 count + Buffer 5 count = 600 bytes
6	256	Frame 3 and end of frame	Buffer 6 count = 256 bytes
7	256	Frame 4	
8	69	Frame 4 and end of frame	Buffer 7 count + Buffer 8 count = 325 bytes
9	256	Frame 5	
10	256	Frame 5 and end of frame	Buffer 9 count + Buffer 10 count = 512 bytes

If there is a possibility that the device can overrun all the buffers, one of two safety mechanisms can be employed. The device can be configured to raise an overrun interrupt, in which case the driver interrupt service routine picks up the received frames. For the second safety mechanism, the driver can poll the controller for frame reception and pick up the received frames. The polling interval can be based on the frame arrival rate (i.e., it is made a function of the buffer-overrun scenarios).

Buffer Handling

Buffer handling depends on the controller. If the controller requires that buffers be located only in a specific area of memory, the driver copies the frames from the driver area to the specified area in memory. It then flags the controller to indicate that the buffers are again available to receive frames. This is a slightly inflexible method but unavoidable due to the restrictions of the controllers. The alternate case is where the controllers can transfer the received frames to any area of memory. So, when a frame is received, the driver removes the linked list of buffers from the controller, hands it a new set of buffers in a different area of memory, and uses the received buffers to pass the frames to its higher layer. This is a more efficient way of handling received frames since there is no performance degradation due to copying frames from the controller buffers to the driver buffers.

Handling Received Frames

The driver passes the received frames to the higher layer. This depends upon the system architecture, i.e., whether the driver and its higher layer are in the same memory space, a common scenario in embedded systems. In such a scenario, the received frame is typically enqueued to the higher layer module by the driver without copying the frame into the module's buffers. An event notifies the higher layer module of the presence of the received frame, so that it can start processing the frame. If more than one frame is enqueued, the driver places all the frames in a First In First Out (FIFO) queue for the higher layer module to pick up.

If the driver and the higher layer are in two separate memory areas, the driver copies the frame into a common area or into the buffers of a system facility like an Inter-Process Communication (IPC) queue and signals the higher layer. The higher layer then copies the frame from system buffers into its own buffers. This approach uses an extra copy cycle—which can degrade performance.

Frame Transmission

Frame transmission also depends on memory and controller functions. The considerations are much the same—whether the memory areas are separate, whether the controller can work with buffers in multiple memory areas, and so on.

In addition, the completion of transmission needs to be handled. The driver can either poll the device or be interrupted on the completion of frame transmission to release the buffers with the data that was transmitted. When there are transmission errors, the driver may need to reset the controller to overcome the errors. To accomplish this, it is important that transmission errors are notified through interrupts as soon as possible. In fact, it is quite common for interrupts to be used only for transmission errors while a regular polling process is used for processing successful transmission completions.

System Issues Related to Drivers

In most systems, drivers are usually responsible for more than one hardware port. The driver differentiates the ports by using data structures—one for each of the ports it handles. These data structures typically involve the memory or I/O address of the controller, the port number, statistics related to the driver, and so on.

A driver may also be a task or a module with no independent thread of execution, since a driver does not exist without a higher layer or lower layer (controller ISR) providing it an impetus to run. A driver handles transmission, reception, and errors, which are all driven by the higher and lower layers. Therefore, many systems implement drivers as either libraries or functions that can be called from the higher layer or from an interrupt service routine.

The alternative, in which the driver is itself a task, allows drivers to implement logic that is useful with hardware controllers. The polling logic for reception is one such case.

If the driver is scheduled as a task that polls the controllers at periodic intervals, we can avoid frame reception overrun. Another use of a separate driver task is when chip statistics counters need to be read within a time interval. It is not always possible to have on chip statistics counters that will never overflow. This is especially the case at higher speeds. A driver task that periodically polls the controllers to read the current value of the counters and maintains them in software will alleviate this situation.

3.4.2 Protocol Functionality

Control Tasks

Figure 3.3 showed the typical module partitioning of protocols in a Layer 2 switch. The switch runs the 802.1D spanning tree algorithm and protocol (STP), which detects loops in the switching topology and permits some ports to be deactivated to avoid the loops. STP frames need to be sent and received periodically by each switch. Transmission of STP frames is initiated by a timer that is maintained by the STP task.

Another protocol used in Layer 2 switching is the IEEE 802.1Q Generic VLAN Registration Protocol (GVRP). A VLAN (Virtual LAN) is a logical partitioninig of the switching topology. Nodes on a VLAN can communicate with one other without going through a router.. Nodes connected to multiple physical LANs (and switches) can be configured to be members of the same VLAN. The switches need to be aware of ports/node VLAN membership and need to communicate this information with each other. GVRP, as defined in IEEE 802.1Q, provides this mechanism. We use the term *control tasks* to describe the STP and GVRP tasks, since these operate in the control plane.

Another method of partitioning could be to combine all the control protocols so that they are handled within one task. This has the advantage of avoiding context switches and memory requirements due to a large number of control tasks. The flip side to this is the complexity—the STP processing may hold up the GVRP processing. If they are separate, equal-priority tasks, equal-priority time slicing could be used to ensure that no one control task holds up the others.

Switching Task

Other than the protocols, there is a switching task that picks up the frames from one Ethernet port and switches them to another port based on the destination address in the frame. The switching task uses the information from frames to build its forwarding table and qualifies the table entries with information provided by the STP and the GVRP tasks. For example, it will not poll the Ethernet driver for frames from deactivated ports (as specified by the STP task). Similarly, it will forward a frame to a port based on the VLAN information it obtained from the GVRP task.

Note that the switching task needs to runs more often, since it processes frames from multiple Ethernet ports arriving at a rapid rate. Due to the nature of the protocols, the STP and the GVRP tasks do not need to process frames as often as the switching task—the control frames associated with these protocols are exchanged only once every few seconds. The switching task thus runs at a higher priority than the other protocol tasks in line with this requirement. If the driver is implemented as a separate task, it needs to have a higher priority than all the other tasks in the system since it needs to process frames as fast as possible. This logic extends upwards also, for the switching task that processes all the frames provided by the driver. The partitioning is done based on the protocol and how often the protocol needs to process frames.

Demultiplexing

Frames are provided to the GVRP, STP or IP tasks through the use of a demultiplexing (demux) operation, which is usually implemented at a layer above the driver. Demultiplexing involves pre-processing arriving frames from Ethernet ports and sending them to the appropriate task. For example, an STP multicast frame is identified by its multicast destination address and sent to the STP task Similarly, an IP packet destined to the router (classified by the protocol type 0x0800 in the type field of the Ethernet frame) is sent to the IP task. With the Layer 2 switch, we assume that the switching task provides the demux function for all frames sent to it by the driver.

Listing 3.1   Perform demultiplexing.

```
{

    If frame is a multicast frame {
        Check destination multicast address
              and send to GVRP or STP task;
    } else
        Dropframe;
        If frame is destined to switch with IP protocol type
          Send to IP function

}
```

In some implementations, the driver can perform the demultiplexing function instead of the switching task—however, it is best to leave this to a separate module layered above the drivers, since there may be several drivers in the system.

In the above example, the algorithm for the switching, i.e., the bridging operation, is not shown. The algorithm includes learning from the source MAC address, filtering, and forwarding of received frames.

TCP/IP End Node Functionality

Layer 2 switches usually have a full TCP/IP stack to handle the following:

- TCP over IP for telnet and HTTP over TCP for switch management
- SNMP over UDP over IP, for switch management
- ICMP functionality such as ping

This implies that the complete suite of IP, TCP, UDP, HTTP, SNMP protocols needs to be supported in the Layer 2 switch. Note that, since there is no IP forwarding performed, the TCP/IP stack implements only end-node functionality. Network managers connect to the switch using the IP address of any of the Ethernet ports. Figure 3.3 showed a Layer 2 Ethernet switch with complete end node functionality.

Often, the TCP/IP end node functionality is implemented with fewer tasks. For instance, IP, ICMP, UDP, and TCP can be provided in the same task. Since end node functionality is usually not time critical, each protocol function can run sequentially when an IP packet is received.

3.4.3 System and Management Tasks

While protocol tasks form the core function of the system, additional tasks and modules are needed for system operation and management. In the Layer 2 switch, an SNMP agent permits an SNMP manager to control and configure the system. The agent decodes the SNMP PDUs from the manager and performs the requested operation via interaction with the individual protocol and system tasks. More details are provided on this in Chapter 7.

A Health Monitor task ensures that hardware and software are performing correctly. The Health Monitor task can tickle the watchdog timer, and, if the task is not scheduled, the watchdog timer will reset the system. Another use of the Health Monitor is to monitor port status. In this case, the Health Monitor task periodically monitors the status of the ports on the system by interfacing to the controller through the driver. If a port is down, it can pass this status up to the driver; this is then propagated up the layers.

Other tasks or modules relevant to system and management functions can include buffer and timer management, inter-board communication, redundancy management, or shelf management in large hardware systems housing multiple shelves.

3.4.4 Hardware Acceleration

In the Layer 2 switch example, it was assumed that the switching was done in software, even if it was inefficient. In reality, switching is often performed by a switching chipset as detailed in Chapter 2. This switching chipset needs to be programmed for the switching parameters, including VLAN parameters, port priorities, and size of filtering tables—all of which can modify the system architecture.

When hardware acceleration is used in our Layer 2 switch example, the switching task is now responsible for the slow-path processing and for demultiplexing frames arriving from the switch. It also programs the switching hardware based on information it receives from the GVRP and STP tasks. It is recommended that we keep the interface to the hardware in a single task or module like the switching task, instead of allowing the GVRP and STP tasks to program the hardware directly.

3.5 Layer 3 Switch/Router

Another example used often in this book is the Layer 3 switch/router This is termed the IP Switch (IPS) in this book. The device performs Layer 3 switching (i.e., forwarding based on Layer 3 information). Similar to the Layer 2 switch, the three types of tasks in a Layer 3 switch/router are: drivers, a set of modules for the protocols, and a set of modules/tasks for system operation and management. The drivers perform the same type of functionality as in the Layer 2 switch.

IP forwarding functionality is provided through an IP switching task while the control plane routing protocols are implemented as separate tasks (control tasks). The IP switching task interacts with the hardware for fast forwarding functions similar to the switching task in the Layer 2 switch example.

The other tasks include a Routing Information Protocol (RIP) task, an OSPF task, and a Border Gateway Protocol (BGP) task. The RIP task runs on top of UDP (User Datagram Protocol), which could be a part of the IP end-node function. This function could be a part of the IP switching task or designed as a separate task for performance reasons.

TCP is part of the same end node task. BGP interfaces to TCP since it needs to run on top of TCP to communicate with peer routers for exchanging routing information. The routing protocols serve to build a routing/forwarding table which is used by the IP switching task to forward packets received on any of its Ethernet interfaces. The interfaces between the routing tasks and the IP switching tasks in terms of access to the routing table are depicted in Figure 3.5.

The modules/tasks required for system operation are similar to the Layer 2 switch case. This includes management tasks like an SNMP agent and Health Monitor task.

The Layer 3 switch operations are more complex than a Layer 2 switch. Protocols like OSPF can consume a substantial amount of CPU time for their Shortest Path First (SPF) calculations, especially when the network is large. Depending upon the application in the network, routers are often built as multi-board systems with a control card and multiple line cards (discussed in Chapter 8).

While Layer 2 switches and IP routers can be separate devices, it is also common for Layer 2 switching functionality to be combined with an IP router, forming a combination Layer 2/3 switch. IP forwarding is frequently called IP switching or Layer 3 switching. This book terms the Layer 2/3 switch as the IP Switch (IPS). Here, the number of tasks/modules will be much higher—requiring a careful design of the system, since inter-task communication can cause the system to slow down considerably.

Figure 3.5   Interface between routing and IP switching tasks.

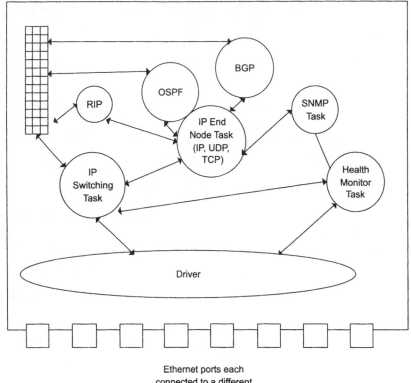

Routing/Forwarding
Table built by
routing protocols
and referred by IP
Switching Task

Ethernet ports each
connected to a different
IP network

3.6 Module and Task Interfaces

Modules and tasks interface with each other using two types of interchange, *data* and *control*. Data is the information to be transmitted or received on an interface. This can include the actual payload packets like packets to be switched, control packets like STP packets, OSPF packets, and management packets like SNMP packets. Control relates to passing control or status information between modules. Control information is totally internal to the communications system—these frames are not transmitted out on the interface.

There are two common schemes of information interchange between tasks/modules:

- Functional or procedural interfaces
- Messaging or event interfaces

3.6.1 Functional/Procedural Interfaces

Procedures or functions are used when two modules are in the same task, or in the same memory context. Consider the Layer 2 switch in which the switching task calls a driver function SendFrame() to transmit a packet. The driver provides this routine as part of a driver interface library so that the higher layer can use it to request the driver to transmit a packet on the Ethernet interface. SendFrame() will execute the steps for queuing the packet to the controller, setting up the transmit registers of the controller, and so on.

The function returns a value indicating the status of the operation. If the function returns immediately after queuing the frame to the controller, the calling function may not know the status of the transmission. Queuing the frames does not mean the transmission was completed. If the function waits for the transmission to complete, the switching task will be held up by the controller transmission status. The function cannot return until the transmission is complete. The call is known as a *synchronous* call, since the function does not return or *blocks* until the complete action is performed.

A blocking API is not a desirable way of implementing the driver, since transmission by the controller can be delayed based on the physical interface, and cause the switching task to be held up. Ethernet's CSMA/CD protocol requires that the transmitter wait for a random period of time if it detects a collision before attempting to retransmit a frame. If the switching task blocks due to this condition, it can delay all other processing.

The alternative is an implementation in which the SendFrame() function returns immediately once the queuing has been done. The status of the transmission is not returned, as it is not complete. The switching task can poll the driver at a later time to find the status. If the transmission has been successful, the buffers can be released via the same polling routine. In fact, the routine is implemented as a driver library and called by the switching task.

The alternative to polling is an interrupt from the driver, but, as discussed earlier, this has its own set of issues. One way to overcome this is with a *callback* function. In this design, the switching task provides a function reference to the driver when it registers itself (see Figure 3.6). The driver calls this function when it receives a transmit completion interrupt from the driver. This is the *asynchronous* mode of operation, with the callback routine indication providing the final status of the transmit operation.

The callback routine, while implemented in the switching task, is called from the Interrupt Service Routine (ISR) by the driver. Though this routine is "owned" by the switching task, it is always called by the driver—a subtle distinction. The switching task's data is visible only inside this routine.

A key requirement of interrupt handlers is that they complete their work and return quickly. Since callback routines can be called from an interrupt handler, it is best to keep the callback routine very short. To fulfill this requirement, the callback routine sends an event or notification to the switching task and exits immediately.

Figure 3.6   Callback function.

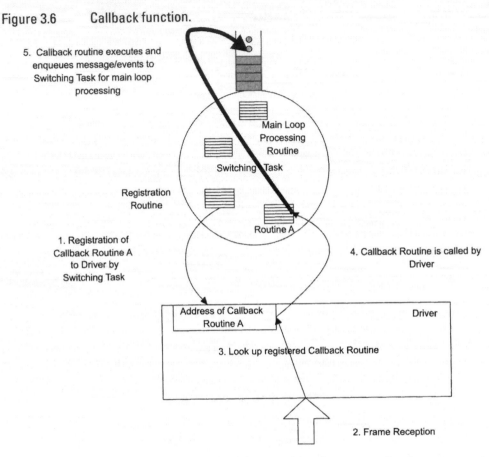

5. Callback routine executes and enqueues message/events to Switching Task for main loop processing

Main Loop Processing Routine

Switching Task

Registration Routine

Routine A

1. Registration of Callback Routine A to Driver by Switching Task

4. Callback Routine is called by Driver

Address of Callback Routine A

Driver

3. Look up registered Callback Routine

2. Frame Reception

The main loop of any task implemented in the communications system waits on events like message reception, timer expiration, inter-task communication, and so on. Each event is identified by its type and additional parameters. On event reception, the task will classify and process the events, as shown in Listing 3.2.

Listing 3.2   Main task loop.

```
Task Main Loop:
    Do Forever {
        /* Hard Wait - call will block if there are no events */
        Wait on Specific Events;
        /* If we are here, one or more events have occurred */
        Process Events;

    }
```

The callback routine event is another type of event, and its parameters are specified by the callback routine when called from the ISR. Instead of one event type per callback function, there is just one callback event, with parameters specifying the type and related information. A callback event with a type indicating transmission completion may result in a list of buffers being released.

This model of notification of the task via events is very useful. It can also be applied to the second method of information interchange, which is the messaging/event interface

3.6.2 Messaging/Event Interfaces

This interface is useful when two tasks in separate memory areas need to communicate. A source task communicates with the destination task by constructing a message and sending it to the destination task. The sending is done with a library call, typically implemented in the operating system or a message library. The message can contain data or control information. The message library copies the data from the source task context into a common area and notifies the destination task—which, in turn, copies the message back into its own memory context.

Even if the two tasks are in the same memory area, this is a useful approach. One way of implementing messaging is with a task-based message queue (see Figure 3.7). Each task has its own message queue and can read and write messages to its own queue, while other tasks are only allowed to write to the queue. In Figure 3.7, a routine in Task 1 makes an interface call to pass some information to Task 2. This call translates into queuing of a message to Task 2. While the module which calls this routine sees only a procedural interface, the implementation of the routine results in a message being sent to Task 2. This is a common technique used in communications software.

The nature of the interface (procedural or messaging) is hidden from the calling routine via an API. Only the implementation of the API will need to use the appropriate interface.

The messaging interface scheme fits in very well with both distributed and multi-board architectures. For example, a switch may be implemented with control tasks such as STP and GVRP on a control card, while the forwarding or switching is done on a line card. The same code from a multi-task environment can be used in a multi-board environment as long as messaging is used between the tasks or if the API is based on messaging.

Some RTOS vendors base their operating systems on message passing. A case in point is the OSE™ real-time operating system, which implements message passing as the only IPC mechanism. The OS also provides a software component called the Link Handler, which helps build distributed systems using messages. Other vendors provide messaging as one of several IPC mechanisms, allowing application engineers to make a choice depending upon their application.

Figure 3.7   Implementing messaging with a message queue.

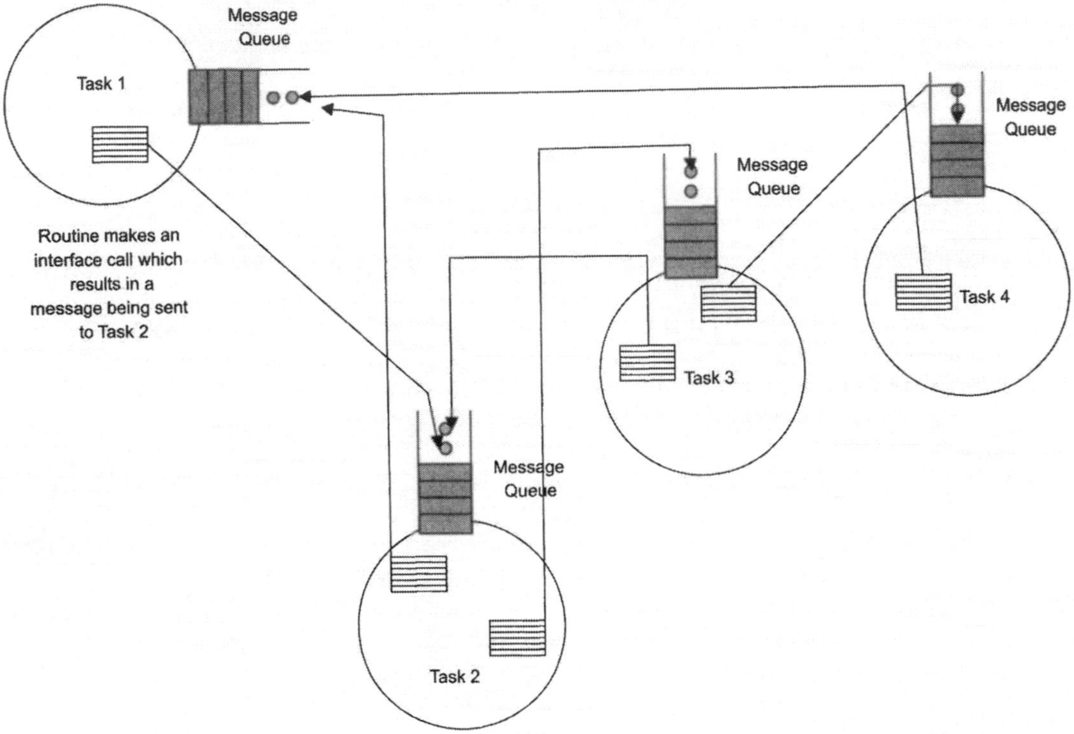

An event can be considered a special type of message—it is sent from one task to another and can include a small amount of additional information which is used by the task. An event can also be an effect caused by message enqueuing or timer expiration. In this context, the main loop is modified to look like Listing 3.3.

Listing 3.3   Event processing in the main loop.

```
While (1)
        Wait for any of the events ( .....0
        /* break out of the hard wait loop */
        if (messageQueuing event)
             Process MessageQueue
        if (timer event)
             Process TimerEvents
```

```
        if (callBack event)
              Process CallBackEvents; /* based on type + params */
        Perform Housekeeping functions;  /* Releasing buffers.. */
    }
```

3.6.3   Standard versus Proprietary Interfaces

Standard operating system calls can be used for Inter-Process Communication (IPC). IPC mechanisms may differ slightly between operating systems, but standards such as POSIX attempt to correct this problem. Software usin : POSIX-compliant API calls is generally portable across multiple operating systems. Th : is true with embedded real-time operating systems as well. General systems calls that originated in the UNIX world such as the socket API, are very popular in the embedded world.

Much of the original TCP/IP networking code came from the 4.2 UNIX BSD (Berkeley Systems Distribution) from the University of California, Berkeley. The socket library and API, in fact, was a 4.2 BSD creation. So applications like telnet and ftp, which came with the 4.2 BSD code, used this API. Some RTOSes used the BSD networking code baseline for their networking stack implementation. This helped them to get to market faster with proven code, thus increasing the prevalence of standard API calls in embedded systems. Also, software engineers moving into embedded software from the desktop communications world found this to be a familiar environment, increasing its use.

The advantage of standard API calls is that the code need not change when moving to another operating system providing the same APIs. A key factor in embedded communications systems is the portability of protocols as an application migrates from one platform to another. As long as the new OS supports the same set of system APIs, the system interface does not change.

We use the term *proprietary interfaces* to imply those interfaces which are specific to the system and which do not use standard interfaces such as POSIX. Proprietary is not necessarily a problem. Sometimes, developers do not want to use the standard system calls due to the performance impact. For example, due to the 4.2 BSD legacy, a send call in the socket API results in a copy into the implementation's buffers, even in those RTOSes where there is no concept of user and kernel spaces. Any data copy results in a performance hit and should be avoided.

Code Maintainability

When designing communications software, tradeoffs between performance and maintainability are weighed. *However, on the first attempt, engineers should focus on code maintainability and functionality with some overall performance goals.* Avoiding copying of messages, reducing the amount of time spent in interrupt handlers, or larger polling intervals, are some of the techniques used to increase performance. After these are followed, engineers should use optimizing compilers as much as possible before attempting to optimize specific sections of their code.

Code tends to survive long after a project has ended. Therefore, it is very important that code be standard and maintainable so that changes can be made easily during migrations to new hardware platforms. The maintainability helps successive generations of engineers understand how the application was designed.

3.7 Summary

The OSI model is useful for layered development, but various factors like protocol dependencies and performance require the developer to deviate from strict layering. The system decomposition involves breaking up the functionality into modules and tasks. A module can be implemented as multiple tasks, while a task may comprise multiple modules.

The key modules in a communications system are the drivers, protocol modules, and the management/housekeeping modules. The drivers will need to interface with the hardware controllers, including the use of hardware acceleration where appropriate. Protocol modules can be combined into one task or operate as individual tasks.

The modules interact with each other via functional interfaces when they are in the same memory space and via messages if they are in separate memory spaces.

3.8 For Further Study

Rich Seifert [2000] provides more detail on the design issues involved in a Layer 2 switch. Ganssle [2001] discusses some do's and don'ts for interrupt handlers.

3.9 Exercises

1. Why do you think a pseudo header was defined in TCP and UDP?
2. Develop the logic for a routine which can be called by a task in a non-preemptive system to hand over control to the scheduler. The handover should occur if there are no events to process and/or if the task has determined that it has consumed enough CPU time.
3. Draw an architectural diagram of a driver implementing a device-dependent layer and a device-independent layer. Outline the functions of each layer.
4. Outline all the protocols required for a TCP/IP end node implementation, and show the interaction between them via a diagram.
5. Provide an example of a callback function and the sequence of operations for its invocation. Also, detail the functionality of the callback function.

CHAPTER 4

Protocol Software

Protocols define the common language used for communicating systems. Protocols form the core of the system function and can be defined at each layer in the OSI reference model. In a typical communications system, modules typically support these protocols. This chapter focuses on protocol implementation, including state machines, interfaces and management information.

4.1 Protocol Implementation

Protocols can be defined via standard specifications from bodies such as the ITU-T, IETF, IEEE, ANSI, and so on. Protocols can also be proprietary and defined by a vendor for use in communicating with their own equipment. Proprietary protocols may be required in those cases in which the vendor feels that existing standards-based protocols are insufficient for their application or in cases in which the vendor believes that their proprietary protocol may give them a competitive advantage. In either case, protocols need to be defined in a specification. There are several protocol specification languages and tools such as Specification Description Language (SDL). Independent of the tool/language used, a protocol specification involves:

- The architectural relationship of the communicating entities—for example, a master–slave mode, or peer-to-peer mode
- A set of valid states for each of the communicating entities—for example, initializing, active, or disconnected
- A set of messages called Protocol Data Units (PDUs) used by the communicating entities
- Timers used by the communicating entities and their values
- Actions to be taken on receipt of various messages and events

The communicating entities defined in a protocol specification assume various modes based on the nature of the communication. The specification defines which messages are sent by the communicating entities and in which mode the messages are valid. Master–slave communication, as in IBM's Synchronous Data Link Control (SDLC) protocol uses two modes for the protocol—*master* and *slave*. Telecom and WAN equipment often have a *user mode* and *network mode*. In a protocol like Frame Relay LMI (Local Management Interface), equipment located at a customer premise (termed Customer Premise Equipment or *CPE*) plays the role of the user, while the Frame Relay switch interfacing to it operates in the network mode. Here, the user node queries the network equipment for the status of the network connections on the link through a Status Enquiry message. The network equipment responds with a Status message with this information.

Figure 4.1   A Simple Protocol State Machine.

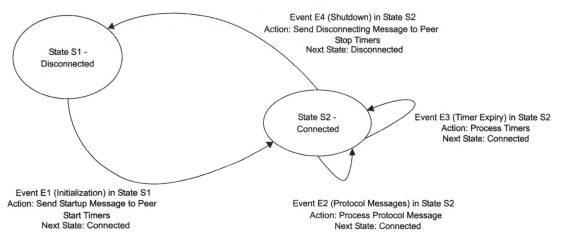

4.1.1   State Machines

Protocols can be either *stateful* or *stateless*. A stateful protocol depends on historical context. This implies that the current state of the protocol depends upon the previous state and the sequence of actions which caused a transition to the current state. TCP is an example of a stateful protoco. A stateless protocol does not need to maintain history. An example of a stateless implementation is IP forwarding, in which the forwarding operation is performed independent of the previous action or packets.

Stateful protocols use constructs called *state machines* (sometimes called Finite State Machines or FSMs) to specify the various states that the protocol can assume, which events are valid for those states, and the action to be taken on specific events. Consider a protocol having two states—Disconnected and Connected (see Figure 4.1). In the Disconnected state, an Initialization event enables the transition to the Connected state. Similarly, valid events in the Connected state are protocol messages and timer events. A

Disable event causes the protocol to move from the Connected state to the Disconnected state. The states and transitions described in Figure 4.1 are very simple. A real protocol implementation has many more states and events.

The implementation of the simplified Connect/Disconnect state machine can be done with a switch and case statement, as shown in Listing 4.1.

Listing 4.1   A simple state machine implementation via a switch-case construct.

```
switch (event) {
    case E1: /* Initialize */
        If (current_state == DISCONNECTED) {
            InitializeProtocol ();
            current_state = CONNECTED;
        }
        break;

    case E2: /* Protocol Messages */
        If (current_state == CONNECTED) {
            ProcessMessages ();

        }
        break;

    case E3: /* Timer Event(s) */
        If (current_state == CONNECTED) {
            ProcessTimers ();

        }
        break;

    case E4: /* Disconnect Event */
        If (current_state == CONNECTED) {
            ShutdownProtocol ();
            current_state = DISCONNECTED;

        }
        break;
```

```
        default:
            logError ("Invalid Event, current_state, event);
            break;


    }
    /* Perform other processing */
```

The previous example is a simple way to implement a State Machine, but it is not very scalable. With several states and events, a switch and case statement would be extremely complex, becoming difficult to implement and maintain.

An alternate method is to use a State Event Table (SET). The concept is quite simple—we construct a matrix with M rows and N columns, in which N represents the number of states, and M represents the number of events. Each column in the table represents a state, while each row represents an event that could occur in any of the states. An individual entry is in the intersection box of the state and the event and represents a tuple—{Action, Next State}, as shown in Table 4.1. For example, the entry at the intersection of S1 and E1 implies:

- The action to be performed on the occurrence of the event E1 while in state S1.
- The next state to transition to on completion of the action—note that it could be the same state, S1 itself.

Using the State Event Table shown in Table 4.1, a typical state machine access function would use the logic in Listing 4.2.

Listing 4.2   Logic for an access function.

```
/*  Entry for current state and event is SET [Event][CurrentState] */


Perform Action (SET [Event][CurrentState]);
CurrentState = Next State (SET [Event][CurrentState]) ;
```

Typical events are specific, pre-defined types of messages, timer events, maximum retransmission attempts, a port going up or down, and error conditions like invalid messages, or user intervention conditions like protocol enabling and disabling.

States depend upon the type of protocol module implemented. Certain events will not be valid in some states. In those cases, the SET entry action would indicate an error. Action routines can be shared because two entries may have the same action.

Table 4.2 depicts the SET for the simple state diagram provided in Figure 4.1.

Table 4.1   State event table.

	State S1	State S2	State S3	State S4
Event E1	{Action, Next State }	{Action, Next State }	{Action, Next State }	{Action, Next State }
Event E2	{Action, Next State }	{Action, Next State }	{Action, Next State }	{Action, Next State }
Event E3	{Action, Next State }	{Action, Next State }	{Action, Next State }	{Action, Next State }
Event E4	{Action, Next State }	{Action, Next State }	{Action, Next State }	{Action, Next State }
Event E5	{Action, Next State }	{Action, Next State }	{Action, Next State }	{Action, Next State }

Table 4.2   SET for simple state machine in Figure 4.1.

	State S1 Disconnected	State S2 Connected
Event E1 Initialize	{{Action: SendStartupMessage, Start Timers}, Next State = S2 }	{{Action: LogError}, Next State = S2 }
Event E2 Protocol Messages	{{Action: LogError}, Next State = S1 }	{{Action: ProcessMessages}, Next State = S2 }
Event E3 Timer Events	{{Action: LogError}, Next State = S1 }	{{Action: ProcessTimers}, Next State = S2 }
Event E4 Disconnect	{{Action: LogError}, Next State = S1 }	{{Action: SendShutdownMessage, Stop Timers}, Next State = S1 }

Actions

If an event is not valid in a state, the protocol implementation can perform one of two actions—a No-Op (i.e., no operation or do nothing) or call an error routine (as in the table above). For example, protocol messages may be received from an external entity even when the protocol is in the disconnected state, which may or may not be considered an error. Good defensive programming identifies all abnormal behavior upfront, before the system is deployed. In that context, it is a good idea to log errors in which unexpected events occur.

Several protocol specifications identify states and events, so it is relatively easy for the communications designer to construct a SET from the specification. Action routines are invoked on specific triggers such as timer expiration and/or received messages. The action routine can cause the construction of the relevant message, after which it can schedule the message for transmission.

Using Predicates

In addition to the two fields in the SET entry, there could be a third field, commonly called a *predicate*, which can be used as a parameter to the action and its value used to decide among various alternatives within the action routine. With predicates, a third entry can be added to {Action, Next State} so that it becomes {Action, Next State, Predicate}. The predicate may also be altered by the state machine. Using the State Event Table (SET) of Table 4.1, a typical state machine access function with a predicate would use the logic in Listing 4.3.

Listing 4.3   Logic for an access function with a predicate.

```
/*  Entry for current state and event is
   SET [Event][CurrentState] */


Perform Action (SET [Event][CurrentState].
              SET  [Event][CurrentState].Predicate);
CurrentState = Next State (SET [Event][CurrentState]) ;
```

State Machine Processing

In Chapter 2, we had outlined the format of the main loop for a typical communications task. With state machines, this would now look like Listing 4.4.

Listing 4.4   Main loop for a typical communications task with state machines.

```
While (1){
    Wait for any of the events;
    /* break out of the hard wait loop */
    if (MessageQueuing event)
        Process MessageQueue;
    If (timerEvent)
        Process TimerEvents
    Perform Housekeeping functions; /* e.g. release
                                    transmit buffers */

}


ProcessMessageQueue ()
{
```

```
    Determine type of message;
    Classify the message and set the event variable;
    Pass event through the SET ; /*state machine access function*/
    }


ProcessTimers ()
{
    Determine attributes of expired timers;
    Classify the timer type and set the event variable;
    Pass event through the SET; /*state machine access function*/
    }
```

The pseudocode given above for the protocol task provides the basis for the event determination. *Events, messages and timeouts received by the protocol task are translated into events for the state machine.*

Multiple State Machines

Protocols do not need to be implemented with a single state machine—in fact, there are often multiple SETs in a protocol implementation. There could be a separate received message state machine, so the only events are incoming messages. There could be a separate state machine for timers, a separate one for port events, and so on. Often, the protocol specification indicates this separation. The OSPF specification in RFC 2328 from the IETF, for example, specifies a neighbor state machine and an interface state machine.

Each of the state machines can be implemented with its own SET. The advantage of this separation is that each SET needs to specify only its relevant set of events and appropriate actions for those events. This modular and distributed approach to SET design contributes to a cleaner system implementation.

SET versus Switch–Case Constructs

The SET implementation is easier to understand than the switch–case statement, since it replaces code complexity with its matrix data structure, ensuring that all states and events are considered up front. In an implementation with only a few valid events in some states, the SET will have a number of actions in whicha call is made to an error routine or a No-op routine and where there is no state change. In these cases, the matrix will typically look like a "sparse matrix," with just a few valid entries. If this had been implemented with a switch–case construct, all invalid events would have been caught with the default case.

The system designer will need to choose between the two approaches for state machine implementation using the constraints for the system being designed. If the state machine is simple and the SET is a sparse matrix, use a switch–case construct. Otherwise, implement the SET with the events and action routines.

4.1.2 Protocol Data Unit (PDU) Processing

Protocol Data Units (PDUs) are generated and received by the protocol implementation. Received PDUs are decoded and classified according to the type of PDU, usually via a type field in the PDU header. Often, it is the lower layer which makes this determination. Consider a TCP packet encapsulated inside an IP packet in an implementation with separate IP and TCP tasks. A TCP packet uses protocol type 6 in the IP header, so the IP task can pass the PDUs with this protocol type to the TCP task. Subsequently, the TCP task looks only at the TCP header for its own processing.

PDU Preprocessing

A PDU is typically *preprocessed* before it is passed to the SET. Preprocessing performs actions such as packet syntax verification and checksum validation. In the TCP example, the checksum is a "ones complement" checksum calculated across the complete PDU. The sender of the PDU calculates and inserts this checksum in the TCP packet.

The receiver TCP implementation recalculates the checksum based on the received PDU and compares it with the checksum inserted by the sender. If the values do not match, the packet is dropped. Checksum calculation is CPU and memory intensive, since it accesses each byte of each packet. Since high I/O throughput is often a real-time requirement, special I/O processors may be added to the hardware configuration to perform TCP checksum calculations.

Events to State Machine

Using preprocessing, the packet type is determined and the appropriate event passed to the SET. Normally, the PDU is not retained once the processing is complete. An exception is a protocol like OSPF, which retains a copy of each Link State Advertisement (LSA) PDU in its LSA database. This is a requirement of the protocol—the OSPF implementation may need to resend one or more received LSAs based on specific conditions, and it needs to have a copy of the PDU for this purpose. In this situation, the OSPF protocol copies the PDU into its own buffers. Alternately, it can retain the PDU buffer and add it to a linked list. This approach avoids copying of the PDU but may involve retaining the same linked buffer management system. The linked buffer scheme may not always be an efficient scheme for data structures to be maintained by the protoco.

PDU Transmission

PDUs are transmitted by the action routines of the SET. For example, timer expiration can cause the appropriate SET action routine to generate a PDU. Second, a received message such as a Frame Relay LMI Status Enquiry (another event to the protocol SET) can cause the generation and transmission of an LMI Status response message. PDU construction is done by the protocol which allocates buffers, fills in the protocol header fields and contents, calculates the checksum, and queues the PDU for transmission or for passing to the lower layer.

4.1.3 Protocol Interfaces

Protocol tasks do not exist or execute in isolation. They need to interface and interact with the other system components in the system environment. These include:

- Real Time Operating System
- Memory Management
- Buffer Management
- Timer Management
- Event Management
- Inter-Process Communication (IPC)
- Driver Components
- Configuration and Control

The RTOS functions as the base platform on which the protocols execute. It is responsible for initializing the protocol task, establishing its context, scheduling it based on priority and readiness, and providing services via system call routines. Each task requires its own stack, which is usually specified at the time of task creation. The RTOS allocates memory for the stack and sets the stack pointer in the protocol task context to point to this allocated area.

Some of the functions of buffer, timer management, and IPC can be libraries in the RTOS, but, for this discussion, they are treated as separate functional entities.

Memory Management

Memory management functions are required for allocating and releasing memory for individual applications by maintaining the memory blocks in the system heap. Calls such as `malloc` and `free` are examples of common memory management functions.

Unlike desktop systems, real-time systems can have multiple memory partitions. Packet buffers can be maintained in DRAM while tables could be maintained in SRAM, and each of these are viewed as separate partitions with their own memory management functions (see Figure 4.2). In the VxWorks™ RTOS, partitions can be created with the `memPartCreate` call. Individual blocks can be created out of these partitions with the routine `memPartAlloc` and released with `memPartFree`. The system calls `malloc` and `free` are actually special cases of `memPartAlloc` and `memPartFree` acting on the system partition, which is the memory partition belonging to the RTOS itself.

Buffer Management

Buffer management includes initialization, allocation, maintenance, and release of buffers used for receiving and transmitting frames to or from physical ports. There can be multiple buffer pools, each consisting of buffers of a specific size.

Figure 4.2   Multiple memory partitions in a communications system.

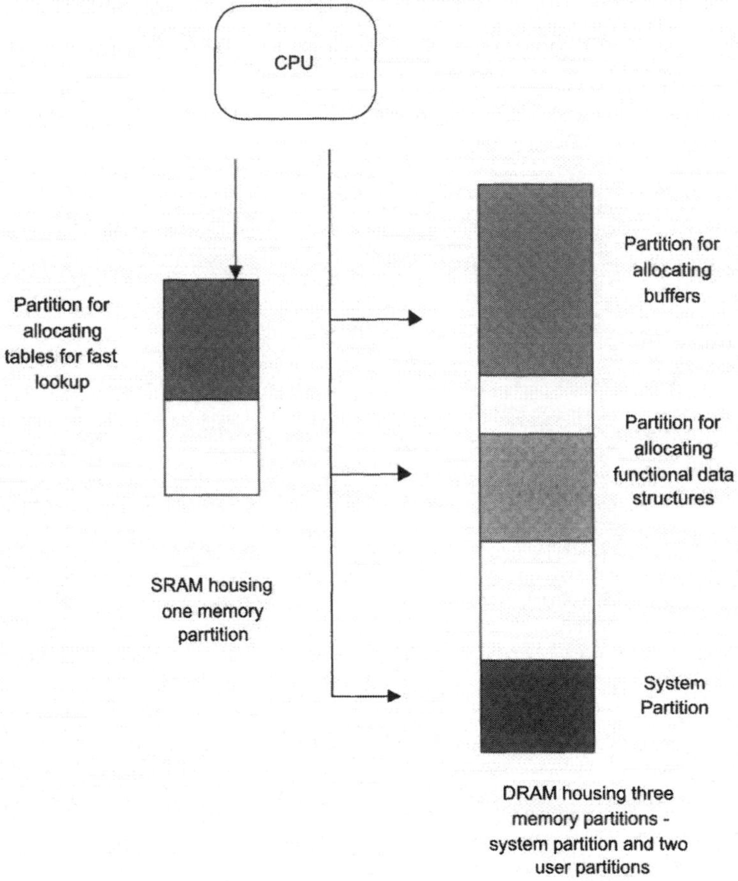

Memory for buffer pool(s) can be allocated using memory management functions. Protocol tasks use the buffer management interface functions to obtain, process, and release buffers needed for their operation. Often, buffer management libraries are provided along with the RTOS—like the mbuf and zbuf libraries available in VxWorks. In some systems, the buffer management library has been developed internally by the software team. This library is considered an "infrastructure" library which can be utilized by all tasks—protocol and otherwise.

Timer Management

Timer management includes the initialization, allocation, management, and use of timers. These functions are provided by a timer management library. As with buffer management, the timer library can either be provided as part of the RTOS or independently developed. Protocol tasks make calls to the timer management library to start and stop timers and are signaled by the timer management subsystem by an event when a timer expires. As indicated in Section 4.1.1, the tasks can use timer expiration as events to the appropriate SET.

Buffer and timer management are discussed in greater detail in Chapter 6.

Event Management

The main loop of the protocol task waits on events. Event management involves the event library, which is used in events such as timer expiration, buffer enqueuing, and so on. The event library also ensures that the task waiting on events is able to selectively determine which signals it will receive. This is usually accomplished using a bit mask that indicates the events that will signal a task. A variation of this is the `select` call used in several RTOSes to wait on a specific set of socket or file descriptors.

Main loop processing of events has the advantage of a single entry point for all events. The main loop can pass control to the appropriate SET based on the type of event. Alternately, if we permit "lateral" calls into the SET, as can happen from an ISR, there would be two issues. The first is that the action routine called from the SET would take a much longer time than is permissible in an ISR. Second, since the SET is a global structure, it would need to use some form of mutual exclusion since there would then be two entry points—one from the main loop and one from an ISR. The preferred approach is for the ISR to send an event to the protocol task, so that event processing and subsequent action routine calls take place only in the main loop.

Inter-Process Communication (IPC)

Tasks use multiple means of communicating with other tasks. These communication mechanisms may be provided by the RTOS or via separate libraries. The mechanisms include:

- Message Queues
- Semaphores for mutual exclusion
- Shared Memory
- Mailboxes (which is a combination of a message queue and a semaphore)
- Signals/Events

These mechanisms are discussed in the literature and will not be detailed here. Selection of one or more forms of IPC depends upon the type of interchange and its availability in the RTOS. Most RTOSes offer these mechanisms—the application developer can choose the appropriate mechanism depending upon the application.

Driver Interfaces

Tasks interface with drivers at the lowest level of the OSI model. For reusability and modularity, a common method for implementing a driver is to segment the driver into two layers:

- An adaptation layer providing a uniform interface to higher layer protocols
- A device-specific layer

The advantage of this layering is that the higher layer tasks need not change with newer versions of the driver. The driver's adaptation layer will provide the same interface to the higher layer/protocol tasks. The device-specific layer and its interface to the adaptation layer will be the only modules which need to be modified. If the next generation of the device and the device driver support more ports, the interfacing tasks will see no difference as long as they deal only with the adaptation layer of the driver.

4.1.4 Configuration and Control

A protocol task communicates with an external manager for configuration, control, status, and statistics. However, the protocol does not talk to the manager directly. It typically interfaces to an *agent* resident on the same embedded communications device. This agent acts on behalf of the external manager (see Figure 4.3) and translates the requests and responses between the protocols and the manager. The manager-to-agent communication is typically through a standard protocol like Simple Network Management Protocol (SNMP), CORBA or TL1. This permits the individual protocols to stay independent of the management protocol and mechanism.

Figure 4.3   A manager–agent model.

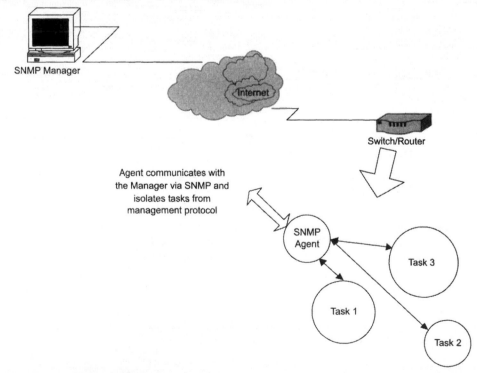

A special case of an agent–manager interaction is the Command Line Interface (CLI). Using the manager–agent model, a CLI can be considered an embedded agent with the user being the manager. Almost every embedded communications device will have a CLI irrespective of whether it has an SNMP agent. The CLI is typically accessible through a serial (console) port or by a remote mechanism such as telnet (when the device implements an end node TCP/IP stack). The user inputs are translated by the CLI task to requests to the individual protocol tasks—very similar to an SNMP agent.

The agent talks to its manager through an external interface, but uses an internal interface to talk to individual protocol tasks. The set of variables to be configured and the status and statistics to be observed are typically defined in a protocol Management Information Base (MIB). The MIB is a database or repository of the management information. Protocol MIBs are usually specified by standards bodies like the IETF.

Management Types

There are four common types of management found in communications architectures:

1. Device or Element Management
2. Network Management
3. Service Management
4. Business Management

The above list details a hierarchy of management functions required in a communications system. Device management feeds into network management, which feeds into service management. A simple way to understand this is with a DSL service example. The service provider can install a DSL modem at the subscriber's location and use SNMP to manage it from a remote location (device management). The information from all modems on the network tells the manager about the status of the network (network management). Service management is one more level of aggregation which helps control the DSL service (downtime, traffic measurement, peering details). Business management determines if the service is making money.

Protocol Management

The following provides a list of some of the operations used in configuring a protocol:

• Enabling and disabling the protocol
• Enabling and disabling the protocol on a specific port
• Addressing a specific interface (e.g., the IP address on a port)
• Setting maximum frame size
• Managing protocol message timeouts
• Timing out peer entities
• Authenticating security information (e.g., passwords, security keys)
• Managing traffic parameters

- Encapsulation information

The set of configuration information is quite extensive, but many parameters have default values specified in the MIBs. Some parameters such as IP addresses do not have default values and need to be set manually. The set of parameters without default values and which require manual configuration are called *basic parameters* in our discussion. These need to be set before the protocol can function. The same consideration applies to individual ports—before a protocol can be enabled on an interface, a set of basic parameters for an individual port needs to be configured. For example, before IP can be enabled on an Ethernet port, the port's IP address needs to be set. While designing protocol software, be sure to identify basic parameters up front—both at the global protocol level and at the port level.

Debugging Protocols

Protocols need to be enabled and disabled before a communications system is supposed to run so that protocols can be debugged. It is useful to isolate the source of error conditions on a network by selectively disabling protocols. A network administrator may want to turn off a specific protocol due to a change in the composition of the network. For example, a network possessing both IP and IPX end stations can transition to become an IP-only network. Instead of replacing the routing equipment on the network, the administrator can just disable IPX forwarding and the related IPX RIP (Routing Information Protocol) and IPX SAP (Service Advertisement Protocol) protocols function on the router.

Once basic parameters are set and the protocol has been enabled, the manager can configure any of the parameters used by the protocol. The manager can also view status of the parameters and tables such as connection tables as well as statistics information.

4.1.5   System Startup

When a communications system starts up, it follows a sequence of steps involving memory allocation, initialization of the various system facilities, and protocols. These steps are performed after the hardware diagnostics are completed and the RTOS is initialized. The latter is required so that the communications software can utilize the OS facilities for its initialization.

The following is the sequence of operations used to initialize a communications system with multiple protocols:

- Initialize memory area and allocate task/heap in various partitions
- Initialize the buffer and timer management modules
- Initialize the driver tasks/modules
- Initialize and start the individual protocol tasks based on the specified priority
- Pass control to the RTOS (which, in turn, passes control to the highest priority task)

There is no step for shutdown or termination above as is often the case in embedded communications systems. There is rarely a reason to shut down the system. Unlike a desktop system, embedded communications devices are in the "always on" mode. Where required, a system reset is done to load a new configuration or image, but this is not the same as a shutdown. This philosophy of "always on" is also seen in the individual protocol tasks, where the main loop is an infinite loop. Tasks wait on events and continue processing without having to break out of the loop for any type of event.

Protocol Initialization

When the protocol obtains control with the initialization of its task, the protocol task performs initialization according to the following steps:

1. Initialize sizing parameters for the tables
2. Allocate memory for dynamic data structures and state table(s)
3. Initialize state table variables
4. Initialize buffer and timer interfaces
5. Read configuration from local source and initialize configuration
6. Initialize lower and higher layer interfaces—including registration with higher and/or lower layers
7. Create and spawn off additional protocol tasks, if required
8. Wait on infinite loop

Setting Sizing Parameters at Startup

This method is preferred for protocol subsystems as opposed to using compile-time constants for the sizing parameters. A compile-time constant requires recompilation of source code when moving to a target with a larger (or lower) amount of memory, so we use variables instead.

These variables can be set by a startup routine, which can read the values from an external entity such as an EEPROM or flash. The sizing variables are then used by the protocol task(s) for allocating the various tables required for operation.

Initialization

The SET initialization is then performed, after which the buffer and timer management interfaces are initialized. This could involve interfacing to the buffer management module for allocating the buffer pool(s) required for the protocol operation. Timers are started as required by interfacing to the timer management module.

Restoring Configuration

Following this is the restoration of configuration—this could be from done a local non-volatile storage such as flash or from a remote host. This is a critical step in the operation of communications devices in general and protocol tasks in particular. Most protocols require an extensive amount of configuration—so when a system restarts due to maintenance, upgrade, or bug fixes, the network manager does not have to reconfigure all the parameters. This is done by saving the system configuration, including protocol configurations, on a local or remote device and having the new image read this configuration at startup.

Application Interface and Task Initialization

The next steps in protocol initialization deal with the registration of modules and callback routines and initialization of interfaces at the higher and lower layers. Subsequently, more protocol tasks may also need to be created. In a router, there could be a root task for TCP/IP subsystem which is responsible for all the protocols in the suite. The root task would initialize and spawn off the additional tasks (IP, TCP, UDP). Each of the tasks can perform its own initialization using the sequence outlined.

4.1.6 Protocol Upgrades

Communications equipment is critical to the functioning of the network. This means that it should not taken out of commission during upgrades, but this is not always possible. In certain environments like the core of the Internet, routers cannot go down for system upgrades. There are several schemes for handling this, including redundancy, protocol task isolation, and control and data plane separation. In particular, protocol task isolation is becoming common in new designs.

Instead of a monolithic design, in which all protocol tasks are linked in with the RTOS and provided as a single image, some vendors are using a user mode—kernel mode design to isolate tasks. The key enablers for this approach are:

- Memory protection between tasks or processes
- Ability to start and stop tasks or processes
- Plane separation between forwarding and routing

Consider a case where an IS–IS routing task runs as a user mode process in UNIX with the actual IP forwarding done in the kernel (with hardware support, as appropriate). If the IS–IS task needs to be upgraded, we can configure the current IS–IS task to save its configuration in a common area (say as a file on the same UNIX system), kill the task and start the "new" (upgraded version) IS–IS task. This upgrade task picks up the configuration from the file and establishes a connection with the IP forwarding in the kernel. Forwarding will continue uninterrupted, and so will the other protocol tasks in the system.

UNIX, Linux, and BSD variants are being used as the operating systems of choice in some of the new communications equipment. In multi-board configurations, several equipment vendors run modified versions of Linux or BSD UNIX on the control blade. These modifications are usually in the area of process scheduling and code optimization in the kernel.

4.2 Summary

Protocols in a system may be standards based or proprietary. They may be stateful or stateless. Stateful protocols are realized using state machines, which may be implemented via switch–case constructs or state event tables. More than one state machine may be present in a protocol implementation.

PDUs are preprocessed and provided as inputs to the state machine. Protocol interfaces can include drivers, buffers, timers, and events along with various forms of IPC. Protocol management includes configuration and control of its management parameters, including saving and restoring of configuration. Upon initialization, protocols will need to use settable sizing variables to avoid recompilation.

4.3 For Further Study

State machines are used in several areas, including hardware design, object modeling, and so on. Martin (1998) provides a good introduction to state machines using UML. Perlman (1999) and Tanenbaum (2002) provide detail about protocol operation, including routing protocols. Stevens (1998) discusses several aspects of networking under UNIX.

4.4 Exercises

1. A hardware port can have two administrative states—disabled and enabled. Operationally, it has two states—up and down, based on the physical connectivity. If the port is disabled, the operational status is designated down. Devise a simple state machine to demonstrate the administrative and operational status of the port.

2. Show the SET and Switch–Case implementation for Problem 1.

3. Outline the steps in PDU preprocessing of an IP packet.

4. A controller supporting 4 Ethernet ports has been replaced by another controller supporting 8 ports. List the device-independent adaptation layer and device-specific layer functions, for applications to migrate to the new driver in a transparent manner.

5. The text discusses use of a UNIX-like environment for protocol restarts. Discuss alternate methods.

CHAPTER 5

Tables and Other Data Structures

Earlier chapters specified that communications software modules use several tables for their operation. One of the functions of control plane software is building tables for data plane operations. This chapter details some of the tables and other data structures typically used in communications systems and discusses the related design aspects. While the term "table" implies a data structure involving contiguous memory, this chapter uses the term to also signify data structures with multiple entries, each of the same base type.

5.1 Tables

Tables are used for storing information referenced by the communications system. In a router, a forwarding table is used by the forwarding software/hardware to forward packets. In a frame relay switch, a connection table can provide the switch map from one Permanent Virtual Circuit (PVC) to another. Global configuration and port-related configuration for a device can be stored in tables. The key issues with tables are:

1. Tables are referenced for both reading and writing. So, both storage and access methods should be optimized for frequent references.

2. Tables can be stored in different parts of memory depending upon their application. For example, forwarding tables can be located in fast SRAM (Static Random Access Memory), while other tables such as configuration and statistics are stored in slower DRAM (Dynamic Random Access Memory).

3. Tables can also be organized according to the access method. For example, an IP forwarding table can be organized in the form of a PATRICIA tree. This structure is commonly used to optimize the access of the entries using the variable-length IP address prefix as an index into the table. A MAC filtering/forwarding table used in a Layer 2 switch often uses a hashing mechanism to store and access the entries.

Hashing yields a fixed-length index into the table and is commonly performed by a bit-level operation on the six-byte destination MAC address in a frame.

5.1.1 Using Tables for Management

Configuration, control, status, and statistics information are four different types of management information that can be represented via tables.

- Configuration—refers to the read–write (or read-only) information used to set the parameters and boundaries for the operation. For example, a password is a configuration parameter.

- Control—indicates read–write information used to change the behavior of the communications software module. For example, enabling or disabling a protocol is treated as control.

- Status—specifies read-only information that provides details about the current state of operation. For example, the operational status of an interface is considered a status variable.

- Statistics—refers to read-only information that the module counts or monitors. For example, a variable that counts the number of packets received by the module is a statistics variable.

In the following discussion, the term "Configuration variables" represents both Configuration and Control variables.

The following sections outline some tables (using illustrative data structures) with information about their allocation, organization, and use.

Management Table Example

Management variables are often defined in a Management Information Base (MIB) which is specified by a body like the IETF. RFC 1213, specified by the IETF, defines an MIB for managing the configuration, control, status, and statistics of a system implementing IP forwarding. The MIB structure defined in this RFC is also known as MIB-II. A system implementing MIB-II can be managed by an external SNMP manager which understands the same MIB.

There are two types of configuration variables —standalone variables and those variables which are part of a table. In MIB-II, the variable ipForwarding indicates a variable which can be used to enable or disable IP forwarding. Standalone variables are also known as *scalar variables* or just *scalars*. The second type of variable is used to construct the elements of a table. These are the fields (or columns) of a table with multiple entries (or rows).

The segment in Listing 5.1 extracted from MIB-II details the IP address information for an interface. The table under consideration is the ipAddr Table. The ipAddrEntry is a row in the ipAddrTable. Each element of the SEQUENCE in ipAddrEntry represents an element in the address entry.

Listing 5.1   Management information base.

```
ipAddrEntry OBJECT-TYPE
    SYNTAX   IpAddrEntry
    ACCESS   not-accessible
    STATUS   mandatory
    DESCRIPTION
            "The addressing information for one of this
             entity's IP addresses."
    INDEX   { ipAdEntAddr }
    ::= { ipAddrTable I }


IpAddrEntry ::=
    SEQUENCE {
        ipAdEntAddr
            IpAddress,         // IP address of interface
        ipAdEntIfIndex
            INTEGER,           //Interface Identifier
        ipAdEntNetMask
            IpAddress,         //Net mask for interface
        ipAdEntBcastAddr
            INTEGER,           //Broadcast address used on interface
        ipAdEntReasmMaxSize
            INTEGER (0..65535) //Max Size of reassembled IP
                                //packet
    }
```

The prefix "ipAdEnt" is used for all the fields of the table. The index to the table is the IP address as specified by ipAdEntAddr. The index will uniquely differentiate the various elements of the table. A tabular representation of the address table, with some entries, is shown in Table 5.1.

Table 5.1   Address table.

Addr	IfIndex	NetMask	BcastAddr	ReasmMaxSize
10.0.0.1	1	255.0.0.0	10.255.255.255	15000
20.0.0.1	2	255.0.0.0	20.255.255.255	18000
30.0.0.1	3	255.0.0.0	30.255.255.255	15000
40.0.0.1	4	255.0.0.0	40.255.255.255	11800

The Addr field is the IP address for the specific interface (identified by an ifIndex). The NetMask, BcastAddr and ReasmMaxSize are configurable fields in a table entry. These variables are also known as "Read–Write" variables as opposed to "Read Only" variables, which cannot be configured by the manager.

The developer can use these definitions for data structures required for the protocol. This can be done via grouping the scalar variables according to configuration, status, and statistics functions, providing a useful segmentation.

Note that the implementation may not follow the exact segmentation and variable set, i.e., the data structures may not be the same as the MIB definition. In the example above, an IP address table implementation may not be a separate table but be a part of an interface or port table. To the SNMP manager, the implementation is transparent, as long as the access and information are returned for the MIB table variables requested.

SNMP MIB tables should be used for completeness. When implementing data structures and their corresponding entries, always use the SNMP MIB tables as a checklist so that none of the variables are missed.

5.2 Partitioning the Structures/Tables

To begin the design of structures and tables, first determine the type of information. Two common types of information are *global* and *per port information*. Global information indicates scalar variables and tables that are global to the protocol or module, independent of the interface it is operating on. Per-port information specifies those scalars and tables that are related to the module or protocol task's operation on a port.

Each protocol or module can have global and per-port information. For example, if IP and IPX run over an interface, each of these protocols will have their own interface-related information, including information like the number of IP or IPX packets received over an interface. Apart from this, each physical interface can have management information that is related to its own operation, such as the port speed or type of physical connection (such as V.35, RS-422). These are maintained by the device driver, which also has its own global and per-port information.

5.2.1 Control Blocks

To design data structures and tables for protocol and system modules, start off with a root data structure called the *Control Block* (CB). This data structure is used to store global information about the module and is used as the initial reference point to access the module's data structures.

Figure 5.1   Physical & Logical Interfaces on a Frame Relay router.

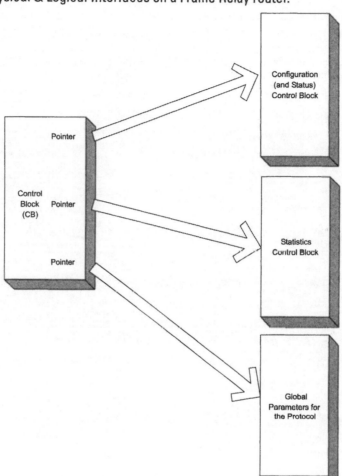

In the case of a protocol, the CB has pointers to the configuration, control, status, and statistics variables for the protocol. These variables themselves can be organized in control blocks (see Figure 5.1). In the figure, we have assumed that the configuration, control, and status variables are available as a single block and that the statistics variables are available as another block. Thus, the Config/Control/Status block (called a Configuration block hereafter) contains Read–Write and Read-Only variables, while the Statistics block contains Read-Only variables.

The CB is the "anchor block" for the protocol from which other information is accessed. The CB can have global information about the protocol itself —this could be in the CB (as shown in Figure 5.1) or accessed via another pointer. The global information specified here includes those parameters that are not configured by an external manager and are to do with the functioning of the protocol within the system. Global information can include housekeeping information (e.g., has the protocol completed initialization, or whether the interfaces to buffer and timer management have succeeded).

A sample of a Protocol Control Block (PCB) and the related Configuration and Statistics Blocks for IP are given in Listing 5.2 (note that only some of the fields are provided).

Listing 5.2   Protocol control block and related blocks for IP.

```
typedef struct ControlBlock {
        BOOLEAN  IPInitialized;
        BOOLEAN  IPBufferInterfaceInitialized;
        BOOLEAN  IPTimerInterfaceIntialized;
        IPConfigBlock *pConfig;
        IPStatsBlock *pStats;
};


typedef struct _IPConfigBlock {
        BOOLEAN ipForwardingEnabled;
        UINT2   u2TTLValue; /* to insert in IP packet */
        UINT2   icmpMask; /* which ICMP messages to
                                        respond to*/

        ........
        ........
} IPConfigBlock;

typedef struct _IPStatsBlock {
        UINT4           ipInReceives;
        UINT4           ipInHdrErrors;
        UINT4           ipInAddrErrors;
        ........

        ........
        UINT4           ipOutDiscards;
        UINT4           ipOutNoRoutes;
        ........
        ........
        ........
        ........
} IPStatsBlock;
```

It may appear that accessing the configuration and statistics with this pointer-based indirection is inherently inefficient. Protocol tasks have to access the configuration block by first accessing the CB and then obtain the configuration block pointer. If all the information were in the control block itself, the CB would be a much bigger structure. While this may not be a problem by itself, segmentation of the blocks and access via pointers can make it easy to save and restore configuration for the protocol. Also, individual protocol configuration and statistics blocks can reside in different parts of memory using a pointer-based design—allowing for a lot more flexibility in memory partitioning.

Design Decisions

This book does not mean to imply that the suggested approaches are the only or best way to design and implement the communications software subsystem. Like a number of design decisions, the developer should choose the approach based on the application. For example, a system which requires the fastest performance for access of data structures would not be a suitable candidate for pointer-based indirection. On the contrary, a system that requires flexibility (even it means a hit on performance) may benefit from some of the approaches suggested here.

There is no one right answer to this. However, some key considerations include code migration and maintenance, portability, and performance.

Logical Interfaces

Each communication protocol or module can be run on multiple interfaces. These interfaces could be physical ports or *logical interfaces*. An example of a logical interface is with a protocol like Frame Relay running PVCs over a single serial interface. Each PVC is treated by the higher layer protocol as though it were a point to point circuit to the peer entity (see Figure 5.2). So, the higher layer would consider each PVC termination as a point-to-point physical interface, effectively yielding multiple logical interfaces over a physical interface.

A protocol needs to have an interface-specific configuration to perform such tasks as enabling or disabling the running of a protocol like OSPF. Similarly, statistics information like the number of OSPF packets received on an interface can be part of (logical) interface-specific statistics. The *Interface Control Block* (ICB) handles this information

5.2.2 Interface Control Blocks

The Interface Control Block (ICB) is similar to the protocol control block. There are two types of ICBs—one for the hardware port and one for the protocol interface. The hardware port ICB, also called the Hardware Interface Control Block (HICB), is a protocol-independent data structure. The HICB represents the configuration, control, and statistics related to the hardware port only. The Protocol Interface Control Block (PICB) represents the parameters for a protocol (configuration, control, status, and statistics) on a specific interface.

Figure 5.2   Logical interface.

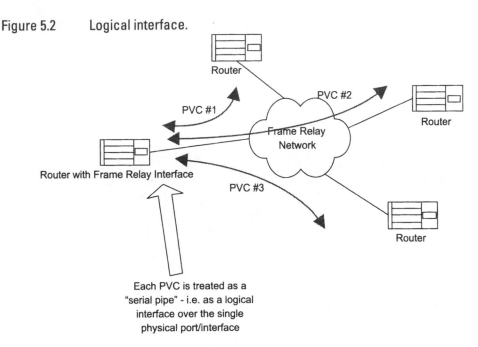

Each PVC is treated as a
"serial pipe" - i.e. as a logical
interface over the single
physical port/interface

Figure 5.3 introduces the HICB and the PICB. *There is one HICB per physical port while PICBs number one per logical interface for the protocol.* Figure 5.3 shows a pointer to the PICB list from the Protocol Control Block (PCB). Each PICB has a pointer to its related HICB and is also linked to the next PICB for the same protocol. Note that the figure shows that PICB 3 and PICB 4 are linked to the same hardware port, HICB 4. They could represent two separate PVCs on a single serial port running Frame Relay, as indicated in Figure 5.2.

Using two types of ICBs rather than a single HICB provides greater flexibility since:

- More than one protocol may be enabled on a hardware port.
- More than one logical interface may be specified on a physical interface.

Consider an example of a multiprotocol router which forwards both IP and IPX traffic over two Ethernet interfaces and a Frame Relay WAN interface. Assume that the Frame Relay port has three PVCs and that IP and IPX forwarding are enabled on all ports and PVCs.

It is clear that both IP and IPX protocols will have their own PCB and PICBs for each of the interfaces they are enabled on. Since there are 3 PVCs on the Frame Relay port, each protocol has a total of five logical interfaces (two for Ethernet, and one for each of the three PVCs on the Frame Relay interface). Each of the PICBs are linked to the HICB for that port. So, the two Ethernet PICBs are linked to the HICBs for Ethernet, and the three PVCs are linked to the HICB for the Frame Relay port. If we had used a single ICB, we would have had to delineate space related to various protocols in the HICB itself—a less-than-ideal situation.

Figure 5.3   Hardware and protocol interface control blocks.

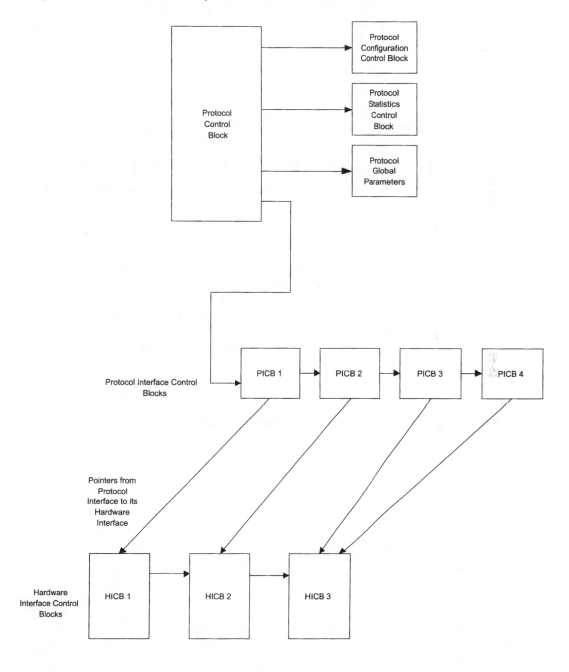

The use of control blocks permits a degree of modularity and stackability (protocols running "on top" of another protocol over a hardware port) of protocol interfaces. In the examples of protocols like IP and IPX running over a serial port, the Frame Relay port is a physical interface. But, Frame Relay is itself a protocol which runs over a hardware interface. So, the FR protocol with its own PCB and PICBs has a one-to-one mapping with HICBs. This is a simple case of "stacking." Configuration of interface stacking is provided via the MIB defined in RFC 1573.

5.3 Implementation

5.3.1 Allocation and Initialization of Control Blocks

A protocol can be enabled or disabled on an interface. But the protocol first requires some basic interface parameters to be set, such as an IP address, before it can be enabled on the interface. This information is usually in the PICB, which needs to be allocated prior to this operation.

A common scheme is to allocate the PICB when, say, an SNMP manager configures the parameters for the protocol on the interface. For example, when a manager sets the IP address for an Ethernet interface, the protocol software allocates a PICB, links it to the Ethernet interface's HICB, and then sets the IP address in the configuration block for the specific PICB. The allocation is done transparently, and the appropriate fields are created and set in the PICB.

As indicated in Chapter 4, non-basic parameters take on default values if they are not explicitly set during PICB creation. If all basic parameters have been set in the PICB, the protocol can be enabled on the interface; if not, the PICB entry is in a "not ready" mode. After all basic parameters have been set, the entry (and PICB) creation is complete.

Allocation Schemes—Static versus Dynamic

ICBs need to be allocated every time an interface is created by the external manager. Software makes a call to the memory management subsystem to allocate the PICB and initializes the fields with values specified by the manager and links to the HICB. The PICB is then linked to the list of PICBs in the PCB.

The advantage of this is that we do not need to allocate memory for the PICB before it is needed. The disadvantage is the overhead of allocating memory on a running system. Note that the peak memory requirement is unchanged independent of when we allocate the memory, as discussed next.

Consider a system where the maximum number of protocol interfaces is 10. Assume that each PICB takes 100 bytes. We contend that the system has to allocate space for 100 × 10 bytes for the PICBs independent of when they are allocated—at startup or during runtime. Other applications should plan on a total memory of 1000 bytes for the PICBs, even if the memory for the interfaces has not been allocated.

Allocation Schemes—Arrays versus Linked List

Memory allocation can follow one of two methods, namely:

1.   Allocate all the PICBs in an array
2.   Allocate memory for PICBs as multiple elements in a free pool list

The array-based allocation is straightforward. Using the numbers above, PICBs are allocated as array elements, each of size 100 for a total of 1000 bytes. A field in each entry indicates whether the element is allocated and provides a next pointer to indicate the next element in the list (see Figure 5.4). Note that all the entries are contiguous.

Figure 5.4   Array-based Allocation for PICBs.

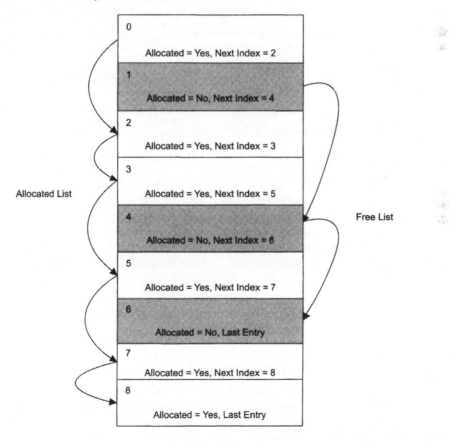

The second type of allocation treats each PICB as a member of a pool. Instead of one large array of 1000 bytes, individual PICBs are allocated using a call such as `malloc` and linked to each other in a free pool. A free-pool pointer indicates the start of the free pool and lists the number of elements available in the pool. Whenever a PICB needs to be obtained by a protocol and linked to the PCB, it is "allocated" out of this free list and linked to the PCB (see Figure 5.5). The free list is empty once all the PICBs are obtained and linked to the PCB. Whenever an interface is deleted with a management command operation, the PICB is "released" back to the free list.

Figure 5.5   Linked List based Allocation for PICBs.

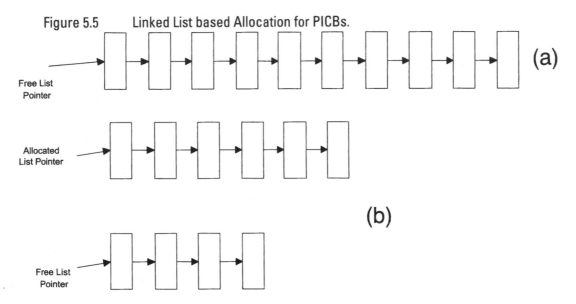

The alternative to a free list is to allocate all the PICBs and link them up to the PCB as they are allocated. An entry in the PCB can indicate the number of allocated and valid PICBs, so that a traversal of the list is done only for the number of entries specified. This method avoids the need to maintain a separate free pool since it can be mapped implicitly from the PICB list itself.

It is best to allocate all required memory for tables and control blocks at startup to avoid the overhead of dynamic allocation during execution.

Applicability to Other Tables

The methods for allocation and initialization of ICBs can be extended to other types of tables and data structures used in communications software. For example, a neighbor list in OSPF, connection blocks in TCP, are data structures where these schemes can be used. In the case of TCP, a connection block could be allocated whenever a connection is initiated from the local TCP implementation or when the implementation receives a connection request from a peer entity. The connection blocks are then organized for efficient access.

5.4 Speeding up Access

Our examples of the PCB and PICB used a simple linked list for organization of the individual elements in the table. This structure, while simple to understand and implement, is not the most efficient for accessing the elements. There are several methods to speed up access based on the type of table or data structure that they are accessing.

There are three ways to speed up access, namely,

1. Optimized Access Methods for specific data structures
2. Hardware support
3. Caching

5.4.1 Optimized Access Methods

For faster access of the entries in a routing table, a *trie* structure has been shown to be more efficient than a linked list. A trie structure permits storing the routing table entries in the leaf nodes of a the data structure, which is accessed through the Longest Prefix Match (LPM) method. Similarly, hashing can be used for efficient storage and access of the elements of a MAC filtering table. The efficiency of the access depends upon the choice of the hashing algorithm and resultant key.

Over Engineering

When planning an embedded application, developers should check prior engineering work for efficient storage and access to data structures. However, there is a risk of over engineering without knowing all the benefits, when developers spend an inordinate amount of time on efficient organization and access, even if the benefits are only incremental. The final goal for an embedded application is a well tested system, with the required features and performance, and within the agreed schedule.

5.4.2 Hardware Support

Another way to speed up access is by using hardware support. An Ethernet controller may have a hardware hashing mechanism for matching the destination MAC address to a bit mask. A bit set in the mask indicates a match of a MAC address that the controller has been programmed to receive.

Another common method of hardware support for table access is with a Content Addressable Memory (CAM). A CAM is a hardware device which can enable parallel searches using a a key. For example, a CAM is often used to obtain the routing table entry corresponding to the Longest Prefix Match of an IP address. This scheme is one of the fastest ways to access and obtain a match—in fact, some modern network processors (e.g., the Intel IXP 2400 and 2800) have built-in CAMs for high-speed access.

> ### A Note on Engineering Assumptions
>
> To illustrate how dynamic engineering assumptions can be, consider the case of a forwarding table used in routers. In the early days of routing, the focus was on reducing the memory requirements for these tables and increasing access speeds through table organization. Due to advances in hardware-based switching techniques and the price of memory dropping considerably, the "traditional" way of organizing the data structures for fast software access were no longer valid. The use of Content-Addressable Memories (CAMs) is a classic example of how this works. CAMs do not require a trie-based organization since the memory is accessed based on content.
>
> Another example of how even protocol focus can change is the example of Multi Protocol Label Switching (MPLS). Originally proposed as means of fixed-header-length label switching (since variable-length longest prefix match or LPM was inefficient for hardware-based switching), the technology shifted its emphasis into traffic engineering (TE). While TE is important, hardware devices became more adept at LPM switching for IP packets, with no discernible drop in performance as compared to a label-switched packet. So, the hardware switching bottleneck became less of an issue, causing the protocol engineers to focus their efforts on the problem of Traffic Engineering.

5.4.3　Caching

Caching is the third method for fast access. A cached subset of a large table can be stored in high-speed memory for faster access. For performance reasons, software always accesses its entries from the cached version. The subset-determination algorithm is key to this scheme, since the idea is to maximize the number of cache hits. If there is a cache miss, the software has to obtain the entries from the lower speed memory, thus negating the effects of caching.

On a Layer 2 switch, the MAC filtering table entries related to the most recently seen destination addresses can be cached. This is based on the premise that the next few frames will also be destined to the same set of MAC addresses, since they are probably part of the same bidirectional traffic flow between the two end stations.

Since the cache size is limited, we need a method to replace entries when a new entry is to be added to a full cache. Timing out entries is one technique. Here, an entry can be removed from the table if it has not been used for a specified period of time. In a Layer 2 switch, a table entry can be timed out if two end stations have not been communicating for a while.

The removal of the entry can be done actively, with a periodic scan of the entries, or it can be done on demand. In the latter case, an entry is chosen for replacement whenever the table is full and a new entry needs to be added. The new entry may be added by a software module called the Cache Handler, which can make the determination based on multiple criteria:

- The entry is accessed more than once in the last few transactions

- Static configuration—an entry is either replaceable or locked via manager configuration
- Prioritization of an entry over another

If the cache is full, the choice of the entry to be replaced can be based on the LRU (Least Recently Used) algorithm used by operating systems. In the packet-forwarding case, this could be an address to which packets are no longer forwarded.

5.5 Table Resizing

It is best to maintain the size of tables on a running system. Table sizes are to be specified at startup in the boot configuration parameters. The system software and protocol tasks read this information and allocate the memory required for the tables. Dynamic resizing of tables, i.e., while the system is running, is not recommended. There are two reasons for this: *reference modification* and *peak memory requirements*.

Reference modification refers to a change in pointers during program execution. Pointer change is not a trivial task, especially in those cases where pointer values have been copied into other variables—a strong reason why dynamic resizing should be discouraged.

Consider a table sized 1000 bytes which needs to be resized to 2000 bytes. A logical approach is to simply resize the table by adding more bytes to the tail end of the table. This is usually not possible since the table and other data structures would have been pre-allocated in sequence from the heap, so there would be no free space to resize the table. Consequently, we need to allocate a new table sized 2000 bytes and copy the contents of the old table to this new table. The first table can be deallocated after the copy is done. However, all references to the earlier table through pointers now need to be changed to point to the new table. This is illustrated in Figure 5.6.

The peak memory required to move data from the old to the new table is the second consideration. Consider the case in which there are only 1500 additional bytes available in the system when the resizing is needed. Since the new table needs only 1000 more bytes, there may appear to be no problem. However, the peak memory requirement during the copy operation is 3000 bytes (1000 for the old table and 2000 for the new table), so memory for the new table cannot be allocated, since we have not released the 1000 bytes for the old table. If there is a large number of tables to be resized, this approach soon becomes unmanageable.

In some MIBs, resizing of tables is permitted by setting a size variable with the SNMP manager. However, the MIB usually specifies that new values will take effect only after the next system reset (reboot), i.e., via modification of the boot parameters. Where resizing is required, this is the recommended approach.

Figure 5.6   Reference modification with table resizing.

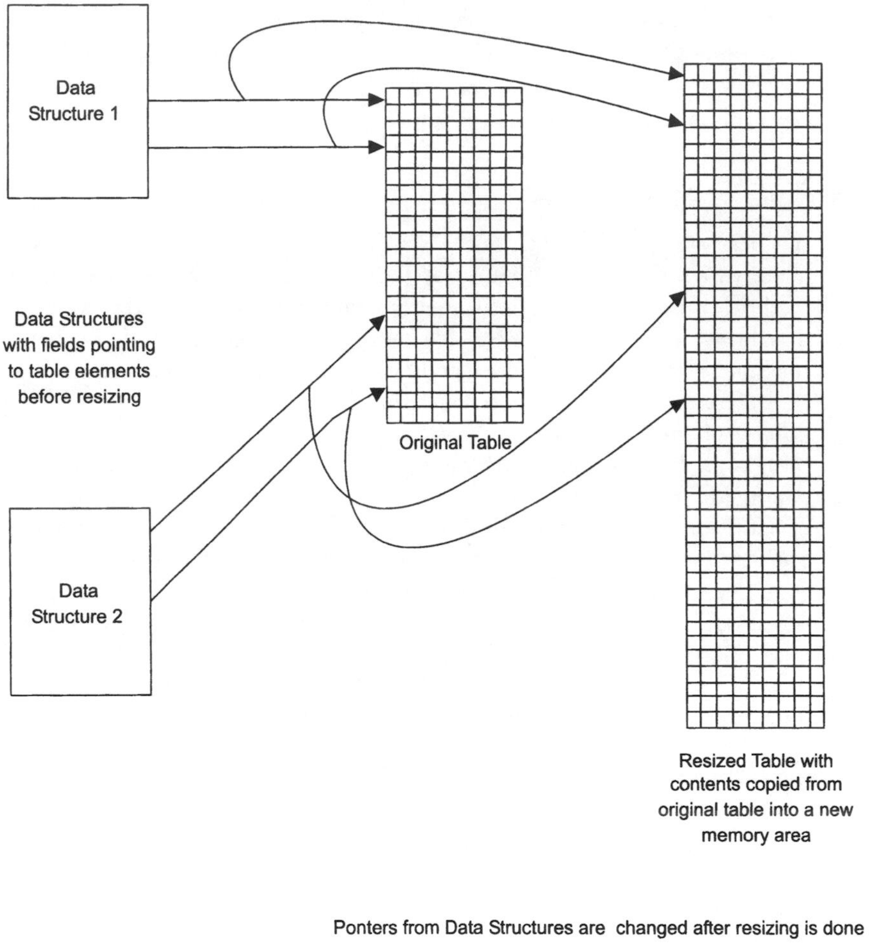

Data Structure 1

Data Structures
with fields pointing
to table elements
before resizing

Original Table

Data
Structure 2

Resized Table with
contents copied from
original table into a new
memory area

Ponters from Data Structures are changed after resizing is done

5.6 Table Access Routines

It is not recommended that tables be accessible by all modules as global entities. The developer should try to avoid directly accessing variables in a global table and instead use access routines that encapsulate the data and functions to manipulate the data into specific modules and submodules. Consider the table shown in Figure 5.7. Instead of directly accessing the table with a pointer pTable, it uses the services of a new module called the *table management module*. This module provides access routines for reading and writing values into this table. External modules will use these access routines only for adding and deleting entries in the table. This concept is quite similar to the encapsulation principles in object-based design.

Figure 5.7 Table access.

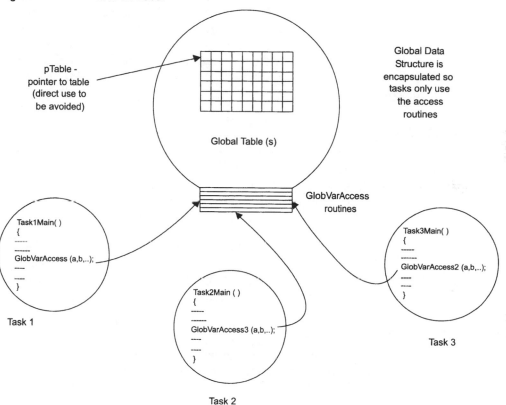

The advantage of using access routines becomes apparent in a distributed environment, where both modules run on separate CPUs but access a common table. The access routine implementations will be modified to accommodate this. Other modules using these routines will not see any difference, since the APIs offered by the access routines will not change. Optimizing the access routines for faster access can also be done in isolation, without having to change the modules that need the access.

Application designers should always use standard access routines for their modularity and ease of maintenance.

If the application always uses access routines to manipulate global or local tables, these access routines should also be written such that critical sections are protected with mechanisms like semaphores. This is to ensure that simultaneous invocation of the routines by two different tasks will not cause unpredictable results.

Another benefit of accessing variables with safe access routines is the ability to write *reentrant code*, a very useful approach in embedded systems. A function is reentrant if it can be invoked by multiple callers at the same time with no side effects on the individual contexts. If variables are changed by the function, the value of the variables needs to be contained within the specific invocation. Reentrant code is useful where multiple contexts of execution are required for the same function(s). Note that a routine is reentrant only if all the routines it calls are reentrant or invocation safe, as in the access routines described here.

5.7 Summary

Tables are used throughout communications software. Tables for management can be constructed out of the standard MIBs for the various protocols and system functions. The Control Block (CB) is the anchor block for a protocol. There is one CB per protocol. The Interface Control Block (ICB) is the per-interface anchor block. There is a Hardware ICB (HICB) per physical port and a Protocol ICB (PICB) for each logical interface for a protocol.

It is best to allocate memory for tables at startup. The allocation can be array or linked list based. To speed up access, efficient table organization and access, hardware support, and caching can be used. Resizing of tables is discouraged. Developers should use table access routines instead of accessing global tables via pointers.

5.8 For Further Study

RFC 1573 provides a good introduction to interface stacking. Comer (2003) details the working and advantages of using CAMs. Jain (2002) provides a comparison of common hashing technique. Ruiz-Sanchez (2001) provides a taxonomy of IP address lookup algorithms. Sridhar (2001) discusses reentrancy in protocol stacks.

5.9 Exercises

1. What are some typical sizes of routing tables in the following: DSL routers, edge routers, and core routers.?

2. Look up the data sheets of popular switching devices (e.g., those from Broadcom, Marvell). Determine the size of the MAC filtering table for Layer 2 tables, and write a note on how it scales as we add more ports (and more switching devices).

3. Construct a spreadsheet for the table requirements for a router. List all the tables with the typical sizes as a function of the number of entries in the table. Modify the sizing parameters to observe how the memory requirements change.

4. List some table access routines for a protocol control block.

CHAPTER 6

Buffer and Timer Management

Buffers are used for the interchange of data among modules in a communications system. Timers are used for keeping track of timeouts for messages to be sent, acknowledgements to be received, as well as for aging out of information in tables. A strategy for Buffer and Timer Management are essential for the communications software subsystem; this chapter covers these two topics in detail.

6.1 Buffer Management

Buffers are used for data interchange among modules in a communications system. The data may be control or payload information and is required for system functioning. For example, when passing data from one process to another, a buffer may be allocated and filled in by the source process and then sent to the destination process. In fact, the buffer scheme in some operating systems evolved from inter-process communications (IPC) mechanisms.

The basic premise of buffer management in communications systems is to minimize data copying. The performance of a system is brought down dramatically if it spends a significant amount of CPU and memory bandwidth in copying data between buffers. The various techniques for buffer management build on this premise.

Buffer management is the provision of a uniform mechanism to allocate, manipulate and free buffers within the system. Allocation involves obtaining the buffer from the global buffer pool. Manipulation includes copying data to a buffer, copying data from a buffer, deleting data from anywhere in the buffer—beginning, middle, or end, concatenating two buffers, duplicating buffers, and so on. Freeing buffers returns the buffers to the global pool so that they can be allocated by other tasks or modules.

6.1.1 Global Buffer Management

Global buffer management uses a single pool for all buffers in the system. This is a common approach in communications systems, where a buffer pool is built out of a pre-designated memory area obtained using partition allocation calls. The number of buffers required in the system is the total of the individual buffer requirements for each of the modules. The advantage of a global pool is that memory management is easier, since the buffer pool size can be increased whenever a new module is added.

The use of a global pool leads to a lack of isolation between modules. An errant or buggy module could deplete the global buffer pool, impacting well-behaved modules. Assume that Modules A, B, and C run three different protocols but use the same global buffer pool. Also, assume that Module A does not release any of the buffers it allocates, thus slowly depleting the buffer pool. Eventually Modules B and C will have their buffer allocations fail and cease operation.

6.1.2 Local Buffer Management

In local buffer management, each module manages its own buffers. The advantage is that buffer representation and handling is independent of the other modules. Consider a module which requires routines only for buffer allocation and release but not other routines such as those for buffer concatenation. In this case, it can have its own "private" buffer management library without the more complex routines. Each module can have the most efficient buffer management library for its operation.

While this provides flexibility, it requires some care at the interface between modules since the representations must be mapped. Moreover, the designer will not have a uniform view of the buffer requirements for the entire system. For these reasons, buffer management libraries are usually global, while buffers themselves can be allocated at either the global or local level.

Third-Party Protocol Libraries

It may not always be possible to design a uniform buffer management library for the entire system. A system may be built with protocol libraries licensed from third-party protocol stack vendors. These vendors could provide their libraries as source or object code. If the libraries are available as only object code, communications system designers do not have visibility into the buffer management scheme of the protocol library. They are aware only of the set of interfaces for data exchange with the protocol module and will use them for buffer interchange.

If the protocol libraries are standard, the interfaces are simplified. For example, the mbuf library in Berkeley UNIX is a global library available to multiple modules. This permits modules to exchange mbufs across their interfaces without the need for mapping.

6.1.3 Single versus Multiple Buffer Pools

Independent of whether we use global or local buffer management, we need to determine the buffer count and buffer size distribution. In a global buffer management scheme, there are two choices:

1. A single set of buffers, all the same size
2. Multiple buffer pools, with all buffers in each pool all the same size

Figure 6.1 illustrates this. In the first case, a single buffer pool is constructed out of the memory area, with the buffers linked to each other. Each buffer in the pool is of the same size (256 bytes). In the second, multiple buffer pools are created out of the memory area, with each buffer pool consisting of buffers of the same size (64, 128, 256 bytes). Note that the size of the memory and the number of buffers are only illustrative—there could be a large memory area segmented into 256-byte buffers or a small memory area segmented into 64- and 128-byte buffers.

Figure 6.1   Single and Multiple Buffer Pools.

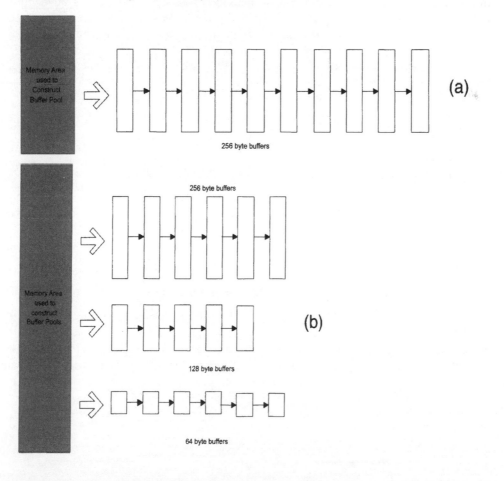

Single buffer pools are easier to manage, while multiple buffer pools can lower the wastage of memory, since the most appropriately sized buffer can be used for the specific frame.

6.1.4 Buffer Size

A rule of thumb for choosing the size of a buffer in the pool is to determine the most common data size to be stored in the buffer. Consider a Layer 2 switch. If the buffers in this device are most commonly used to store minimum-size Ethernet packets (sized 64 bytes), then choose a buffer size of 80 bytes (the extra bytes are for buffer manipulation and passing module information). With this method most frames are sent and received by the device without much buffer space waste. If the frame size exceeds 64 bytes, then multiple buffers are linked to each other in the form of a chain or a linked list to accommodate the additional bytes. The resulting structure is often called a *buffer chain*. Popular buffer schemes like the mbuf library used in Berkeley UNIX follow this format.

If the frame size is less than 64 bytes, there will be *internal fragmentation* in the buffer, a situation familiar to students of memory allocation in operating systems. Internal fragmentation is unused space in a single buffer. When the frame size is larger than 64 bytes, internal fragmentation can occur in the last buffer of the chain if the total frame size is not an exact multiple of 64.

For example, if the received frame size is 300 bytes, the following calculations apply:

Number of buffers required = 300/64 = 4 + 1 = 5 buffers

Size of data in the last buffer = Modulo 300/64 = 44 bytes

Unused data in the last buffer = 64 – 44 = 20 bytes

It is next to impossible to avoid fragmentation in a system if the frame sizes can vary. Designers should instead focus on the optimal size for a buffer in a pool. This size will be the one where the most common frame size will fit without the need to use multiple buffers in a buffer chain.

6.1.5 Checklist for Buffer Pools and Sizes

The following provides a checklist that can be used in selecting a buffer management strategy:

- Use global buffer management if there is no dependency on external modules provided by a third party. Even when such an external module uses its own buffer management, keep a global buffer management strategy for the rest of the system, and define interfaces for clean interchange with the external module.

- If the packet sizes that are to be handled by the system do not vary much, choose a single buffer pool, with an optimal size.
- Avoid buffer chaining as much as possible by choosing a single buffer size closest to the most frequently encountered packet size.

Figure 6.2   The BSD mbuf Structure.

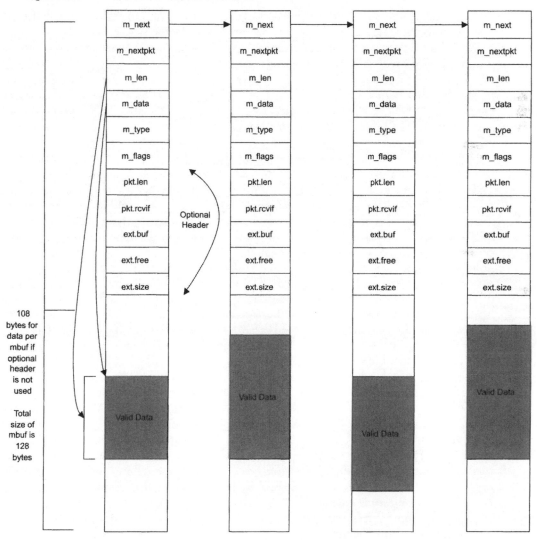

6.1.6   The Berkeley Systems Distribution (BSD) mbuf Library

The BSD mbuf library is discussed in this section to illustrate some buffer management concepts. The BSD mbuf library was first used for communications in the UNIX kernel. The design arose out of the fact that network protocols have different requirements from other parts of the operating system both for peer-to-peer communication and for inter-process communication (IPC). The routines were designed for scatter/gather operations with respect to communications protocols that use headers and trailers prepended or appended to the data buffer. Scatter/gather implies a scheme where the data may be in multiple memory areas or buffers *scattered* in memory, and, to construct the complete packet, the data will need to be *gathered* together.

The mbuf or memory buffer is the key data structure for memory management facilities in the BSD kernel. Each mbuf is 128 bytes long, with 108 bytes used for data (see Figure 6.2). Whenever data is larger than 108 bytes, the application uses a pointer to an external data area called an mbuf *cluster*. Data is stored in the internal data area or external mbuf cluster but never in both areas.

As Figure 6.2 shows, an mbuf can be linked to another mbuf with the m_next pointer. Multiple mbufs linked together constitute a chain, which can be a single message like a TCP packet. Multiple TCP packets can be linked together in a queue using the m_nextpkt field in the mbuf. Each mbuf has a pointer, m_data, indicating the start of "valid" data in the buffer. The m_len field indicates the length of the valid data in the buffer. Data can be deleted at the end of the mbuf by simply decrementing the valid data count. Data can be deleted at the beginning of the mbuf by incrementing the m_data pointer to point to a different part of the buffer as the start of valid data. Consider the case when a packet needs to be passed up from IP to TCP. To do this, we can increment m_data by the size of the IP header so that it then points to the first byte of the TCP header and then decrement m_len by the size of the IP header.

The same mechanism can be used when sending data from TCP to IP. The TCP header can start at a location in the mbuf which permits the IP header to be prepended to the TCP header in the same buffer. This ensures there is no need to copy data to another buffer for the new header(s).

Another significant advantage of mbufs is the ability to link multiple mbufs to a single mbuf cluster (see Figure 6.3). This is useful if the same frame needs to be sent to multiple interfaces. Instead of copying the same frame to all the interfaces, we can allocate mbufs to point to the same mbuf cluster, with a count indicating the number of references to the same area. The *reference counts* are stored in a separate array of counters. Freeing an mbuf decrements the reference count for the corresponding data area, and, when the reference count reaches zero, the data area is released. The mbuf example is an important technique for buffer management and is used in several systems.

The mbuf buffer management scheme is an example of a *two-level hierarchy* for buffer organization. The first level is the mbuf structure, and the second is the mbuf cluster pointed to by the mbuf. Adding data to the beginning or end of the mbuf cluster will require modifying the pointers and counts for valid data in the mbuf.

Figure 6.3 Creating an `mbuf` cluster with multiple `mbuf`s.

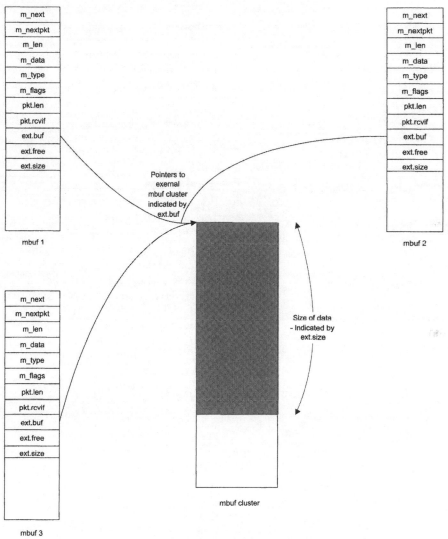

A Quick View of the `mbuf` Library Routines

The routines available in the `mbuf` library include those for allocating a single `mbuf`, freeing an `mbuf`, deleting data from the front or end of the `mbuf`, copying data from an `mbuf` chain into a linear buffer, making a copy of an `mbuf` chain into another, and so on (see Table 6.1).

Table 6.1   mbuf library routines.

Function Name	Description and Use	Comments
m_get	To allocate an mbuf mptr = m_get (wait, type)	wait indicates if the call should block or return immediately if an mbuf is not available. Kernel will allocate the memory for the mbuf using malloc
m_free	To free an mbuf m_free (mptr)	Returns buffer to the kernel pool
m_freem	To free an mbuf chain m_freem (mptr)	Returns buffers to the kernel pool
m_adj	To delete data from the front or end of the mbuf m_adj (mptr, count)	If count is positive, count bytes are deleted from the front of the mbuf. If it is negative, they are deleted from the end of the mbuf.

m_copydata	To copy data from an mbuf into a linear buffer
	m_copydata
	  (mptr, startingOffset,
	  count, bufptr)
		startingOffset indicates the offset from the start of the mbuf from which to copy the data. count indicates the number of bytes to be copied while bufptr indicates the linear buffer into which the data should be copied. We need to use this call when the application interface requires that the contents of the packet be in one contiguous buffer. This will hide the mbuf implementation from the application—a common requirement.

m_copy	To make a copy of an mbuf
	mptr2 =
	  m_copy (mptr1,
	        startingOffset,
	        count)
		mptr2 is the new mbuf chain created with bytes starting from startingOffset and count bytes from the chain pointed to by mptr1. This call is typically used in cases in which we need to make a partial copy of the mbuf for processing by a module independent of the current module.

m_cat	To concatenate two mbuf chains m_cat (mptr1, mptr2)	The chain pointed to by mptr2 is appended to the end of the chain pointed to by mptr1. This is often used in IP reassembly, in which each IP fragment is a separate mbuf chain. Before combining the chains, only the header of the first fragment is retained for the higher layer. The headers and trailers of the other fragments are "shaved" using the m_adj call so that the concatenation can be done without any copying. This is one example of the power and flexibility offered by the mbuf library.

6.1.7 The STREAMS Buffer Scheme

The mbuf scheme forms the basis for a number of buffer management schemes in commercially available RTOSes. An alternate buffer scheme is available in the STREAMS programming model, which was first presented in Chapter 2.

Consider Figure 6.4. which shows the STREAMS buffer organization. There is a *three-level hierarchy* with a message block, data block, and a data buffer. Each message can consist of one or more message blocks. In Figure 6.3, there are two messages, the first having one message block and the second composed of two message blocks.

Figure 6.4   STREAMS buffer organization.

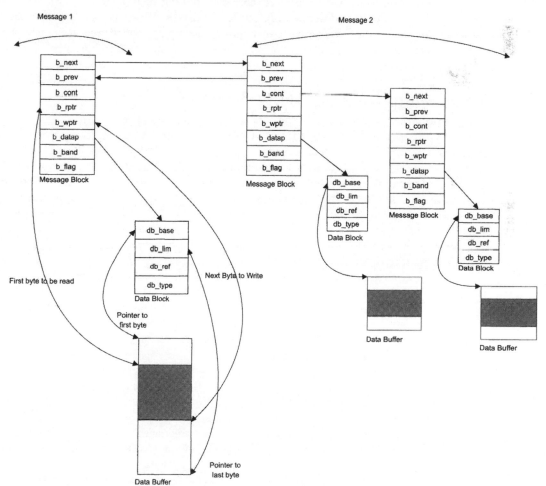

Each message block has multiple fields. The b_next field points to the next message in the queue, while b_prev points to the previous message. b_cont points to the next message block for this message, while b_rptr and b_wptr point to the first unread byte and first byte that can be written in the data buffer. b_datap points to the data block for this message block. Note that the second message has two data blocks, one for each message block in the message.

In the data block, db_base points to the first byte of the buffer, while db_lim points to the last byte. db_ref indicates the reference count, i.e., the number of pointers (from message blocks) pointing to this data block (and buffer).

While the structures may appear different from the mbuf scheme, the fundamentals are the same. The STREAMS buffer scheme uses linking to modify the data without copying, concatenating, and duplicating buffers, and uses reference counts when multiple structures access the same data area. Similar to the separate mbuf table for cluster reference counts, the STREAMS buffer scheme uses the db_ref field in the data block to indicate the reference count for the memory area.

6.1.8 Comparing the Buffer Schemes

The two popular schemes for buffer and chain buffer management are the two-level hierarchy (as in mbufs) and the STREAMS three-level hierarchy. The two schemes are shown in Figure 6.5.

Which of the two schemes is more efficient? The two-level hierarchy is a simple scheme and has only one level of indirection to get data from the mbuf to the mbuf cluster or data area. The three-level hierarchy requires an additional level of indirection from the message block to the data block and to the corresponding data area. This is required only for the first data block since the message block only links to the first data block. The three-level hierarchy also requires additional memory for the message blocks, which are not present in the two-level hierarchy.

In a three-level hierarchy, the message pointer does not need to change to add data at the beginning of the message. The message block now points to a new data block with the additional bytes. This is transparent to the application since it continues to use the same pointer for the message block. With a two-level hierarchy, this could involve allocating a new mbuf at the head of the mbuf chain and ensuring that applications use the new pointer for the start of the message.

The two-level hierarchy is the same as the three-level hierarchy, but the message block is merged into the first data block (or mbuf). Both schemes are used in commercial systems and use an external data area to house the data in the buffer. This is a flexible method for handling data, since it can be used across protocols using different buffer management schemes. Consider a system implemented with protocol stack products from multiple vendors. If each of the products has its own buffer management scheme, we can still provide for data interchange on the interface without any copying if the final data is housed externally. For illustration, we can consider the two stacks to implement a three-level and a two-level hierarchy. Revisiting Figure 6.5, we can deduce how data can be manipulated across the interface; the data area is not copied between the two types of buffer schemes. Rather, only the pointers change.

Figure 6.5   (a)Three and (b) Two level buffer Management Schemes.

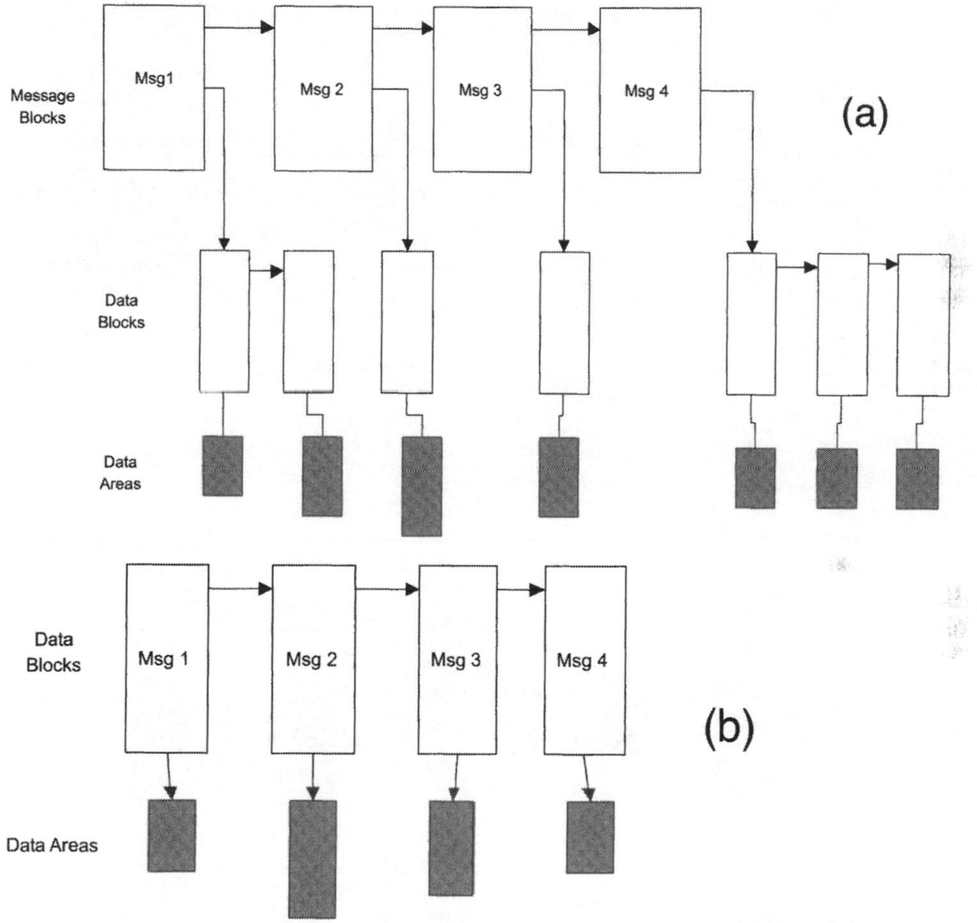

6.1.9   A Sample Buffer Management Scheme

This section outlines the important components of a buffer management scheme using the ideas discussed earlier. In this real target system example, there are three types of structures—a message block, data block, and data buffer (see Figure 6.6). This scheme is similar to the buffer structure in STREAMS implementations.

The message block contains a pointer to the first data block of the message, and the data block contains a pointer to the actual data associated with the block. Message blocks and data blocks are allocated from DRAM and are housed in their own free pools.

Figure 6.6   Structures in a buffer management scheme.

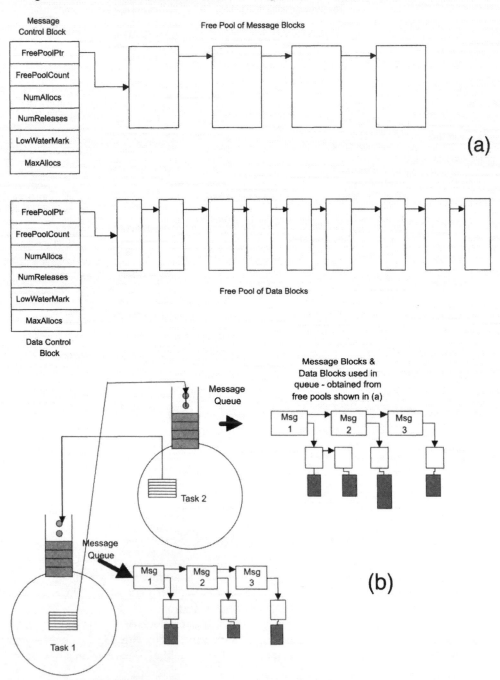

There is a message control block (MCB) and a data control block (DCB) which has the configuration, status, and statistics for the message and data blocks (see Figure 6.6(a)). The buffers should be allocated from DRAM and linked to the data blocks as required while the system is running. Figure 6.6(b) shows the system with two tasks after allocating and queuing messages on the task message queues. As seen, the message blocks maintain the semantics of the message queue.

Data blocks can be used for duplicating data buffers without copying. For example, two data blocks can point to the same data buffer if they need to have the same data content. Routines in the buffer management library perform the following actions:

- Allocating and freeing of data blocks
- Linking a data buffer to a data block
- Queuing messages
- Concatenating messages
- Changing the data block pointer

The library is used by various applications to manipulate buffers for data interchange. One important factor in this buffer management scheme is the *minimization of data copying*—realized by the linking to data blocks.

Message and Data Buffer Control Blocks

The structure below shows the typical format of the message control block. There is a pointer to the start of the free pool housing the available message blocks. The count of available message blocks in the free pool is the difference between the number of allocations and the number of releases (NumAllocs - NumReleases). For the example, assume this is a separate field in the structure (FreePoolCount) (Listing 6.1).

Listing 6.1   Message control block.

```
typedef struct {
    struct MsgBlock *FreePoolPtr;
        unsigned long    FreePoolCount;
        unsigned long    NumAllocs;
        unsigned long    NumReleases;
        unsigned long    LowWaterMark;
        unsigned long    MaxAllocs;
    } MsgControlBlock;
```

When the system is in an idle or lightly loaded state, the free-pool count has a value in a small range. In an end node TCP/IP implementation, which uses messages between the layers and with applications, a message and its message block will be processed quickly and released to the free-pool. Allocations will be matched by releases over a period of time. The difference, i.e., the free-pool count, will not vary much because few messages will be held in the system waiting processing. Sampling the number of queued messages is a quick way to check the health of the system.

When the system is heavily loaded, messages may not be processed and released rapidly, so the free-pool count may dip to a low value. However, when the system comes out of this state, the free-pool count will return to the normal range.

The LowWaterMark can be used to indicate when the free-pool count is approaching a dangerously low number. It is a configurable parameter which is used to indicate when an alert will be sent to the system operator due to a potential depletion of buffers. The alert is sent when the free-pool count reaches a value equal to or below LowWaterMark.

This variable should be set to a value high enough so that the system can allocate buffers to send the alert to the manager about the depletion, permitting the manager to take appropriate action. The depletion may be a temporary phenomenon or could happen when some module is holding up a number of messages. The management action for this alert could be to increase the number of message blocks at the next startup or to shut down the errant module. Choosing the correct value for LowWaterMark can permit a graceful shutdown of the system by the manager.

Similar to the message control block, we have a control block for . The structure is shown in Listing 6.2.

Listing 6.2   Data control block.

```
typedef struct {
struct  DataBlock *FreePoolPtr;
            unsigned long     FreePoolCount;
            unsigned long     NumAllocs;
            unsigned long     NumReleases;
            unsigned long     LowWaterMark;
            unsigned long     MaxAllocs;
        } DataControlBlock;
```

6.1.10 Exception Conditions in Buffer Management

If the system does not have adequate memory for buffers, or if there are issues in passing buffers between modules, the designer would have to provide for exception conditions such as the following:

- Lack of buffers or message or data blocks
- Modules unable to process messages fast enough
- Errant modules not releasing buffers
- System unable to keep up with data rates

The lack of buffers or message or data blocks was covered earlier in the text where we specified using a low-water mark to alert the operator. Designers should engineer pool counts for fully loaded systems, which can be verified when the system is tested in the lab at peak load. It is usually not possible to fix this problem on a real-time system in the field. However, the alert helps the operator determine problems that should be addressed on the next system reboot.

Modules may not be able to keep up with messages for a variety of reasons. They could be spending a lot of time in algorithmic processing, or they may be scheduled less often due to lower task priorities. This causes some modules to hold up buffers. Modules which queue these buffers should perform flow control whenever the destination module queues reach a *high-water mark*. This is a threshold parameter to indicate a safe queue depth or count. The queuing call made by the source module checks if the queue of the target module has crossed the high-water mark. If so, it aborts the queuing operation and returns an error code to the source module. The source module can report this error or retry its operation at a later stage by storing the message in its own queue.

Errant modules are those that are unreliable due to bugs or faulty design. These cause the system to run out of message blocks and buffers. The low-water mark and alert is one way to inform the operator about these modules.

The final type of error is that the system cannot keep up with data rates and runs out of buffers. There are several methods to handle congestion, including well-known techniques such as RED (Random Early Detection/Discard) and Weighted RED. Several of these techniques are now implemented in hardware controllers also.

6.2 Timer Management

Timers are used for a number of purposes in communications systems. There are at least three significant uses for timers, namely:

1. Protocol tasks and system tasks need to periodically perform certain functions. A protocol like frame relay requires a periodic status request message to be sent to the peer. System tasks can also periodically monitor the status of hardware ports using timers.

2. Peers may need to time out based on receiving or not receiving messages. If the peer OSPF task has shut down, it will stop sending "Hello" protocol messages. The local OSPF task will "time out" the peer since it has not received the message in a specific period of time. This will cause a recalculation of the network topology since the peer router is not a part of the network any more.

3. The protocol and system tasks may need to use one-shot timers, which "fire" when the specified time elapses. This can happen when one task contacts another with a specific request. If the second task has not responded within a specified time (as indicated by the timeout), the first task can initiate some error recovery action.

A protocol task may need multiple timers. For example, a task may need to send out a message every 30 seconds and also time out a neighbor if more than 180 seconds has elapsed since it received the last message. As shown in Figure 6.7, the timers may need to be fired at different stages:

Figure 6.7   Managing multiple timers.

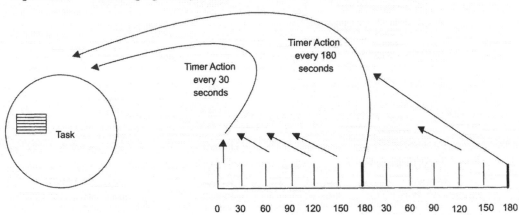

6.2.1 Timer Management Per Task

The example used in Figure 6.7 assumed that timers need to be fired every 30 seconds and every 180 seconds. Assume that the lowest granularity of timers that our system requires is in seconds and that the timer tick provided by the system clock is 10 milliseconds. So, the system requires timer processing once every 100 ticks (100 * 10 milliseconds is 1 second).

The RTOS typically offers some timer-related facilities to manipulate timers based on the system clock, which measures time in clock *ticks*. A board support package (BSP) will hook up to the hardware timer on the board so that it will invoke a timer interrupt on each tick. Most real-time operating systems as well as UNIX systems that support the real-time extensions to POSIX provide facilities for an application to connect to the system clock routine.

The typical sequence of operations for timer ticks in an RTOS environment is shown in Listing 6.3. AppClock () is a routine which is called from the timer interrupt on each tick. Each application can be designed such that it receives a notification after several system ticks. In the example below, App1 requires a notification every 1 second or 100 system ticks. This is realized by incrementing an application tick count (App1Time) and notifying the application App1 when the count reaches 100.

Listing 6.3   RTOS notification routine.

```
AppClock ( )
{
------..
------..
static unsigned long App1Time = 0;
```

```
/* Increment the count every tick and reset when 100 ticks i.e. 1 second
expires */
    Appl1Time++;
    If (Appl1Time == 100) {
        Appl1Time = 0;
        Appl1TimerTickNotify ( );
    }
}
```

The `Appl1TimerTickNotify` function is the timer tick routine for the application `Appl1`.
Figure 6.8 shows how the timers can be organized. A table stores the current timer count
and the context. The table is populated based on the timers that are required dynamically, i.e., during runtime. Figure 6.8 also shows the context, including routines to be
called on timeout (also known as timeout routines), such as an update routine or a
neighbor timeout routine. The context also includes parameters required by these routines. The timeout occurs when the current timer count reaches 0 for a table entry. If it is
a one-shot timer, the entry is removed from the table. If it is a continuous timer, the value
is reset to the initial timeout value (30 seconds and 180 seconds).

Figure 6.8   Table based timer organization.

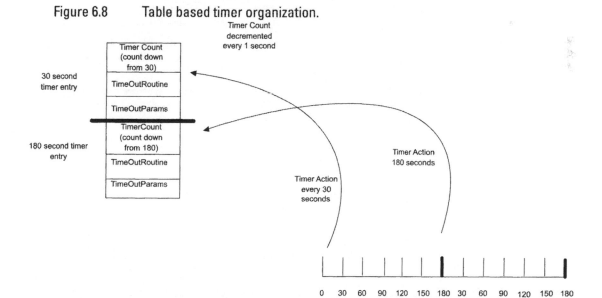

The standard disadvantage of a table-based construct applies to timer functions.
Tables are cumbersome data structures when the application requires a significant
amount of dynamism. The alternative is to use linked lists. The list will consist of entries

similar to the table—however, they will be added and removed dynamically based on the application's requirement.

The term *timer block* refers to each entry in the list. Timer blocks are typically allocated and assigned to a free pool, similar to buffers. To start a timer, a task allocates a timer block from the free pool and links it into a list. The timer block has a count field to indicate the number of timer expirations. As with the table-based approach, a timer tick decrements the timer count in each of the timer blocks until the value reaches zero. At this point, the timeout routine in the timer block is called for each of the entries.

This is a simple approach to address the dynamic requirements for timers during runtime. The only overhead is the need to decrement the timer count in each timer block, a situation compounded by the overhead of pointer dereferencing for linked lists. The solution is to use differential timer counts, as outlined next.

Figure 6.9   Differential timers.

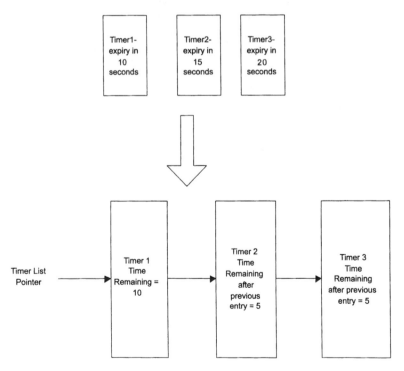

Timer Blocks with Differential Timer Counts

6.2.2 Using a Differential Timer Count

The inefficiency of the multiple decrements is addressed by a differential timeout scheme, which is common in communications software systems. With a differential timeout, the lowest timer count entry is stored at the head of the list. Figure 6.9 illustrates three timer blocks with 10-, 15-, and 20-second timeouts. Instead of storing them in the same sequence, there are three entries, with the first entry's initial timer count set to 10, the second to 5 (since this entry needs a timeout 5 seconds after the first entry's timeout) and the third entry to 5 (i.e., 5 seconds after the second entry's timeout). When timers have the same value, the differential count is 0, and the timeout happens at the same time as the previous entry's timeout. The advantage of this approach is that the current timer count is decremented only in the first entry of the list. When it reaches 0, that entry is processed along with all subsequent entries in which the current timer count is 0. These entries can now be removed from the list.

A typical application using the timer tick notification and the addition and deletion of timer block entries in the timer list is shown below:

- Application calls StartTimer (10, Parameter List ...);
- Application calls StartTimer (15, Parameter List ...)
- Application calls StartTimer (20, Parameter List ...)

These three calls create three separate timer blocks and store the parameter list (context) along with the differential count—which will resemble the organization of timers in Figure 6.9. Each timer block is as in Listing 6.4.

Listing 6.4   Timer block.

```
Struct {
        unsigned long Count;

        ......
        Param1
        Param 2
        ......
        Param n;
} TimerBlockType;
```

TimerTickAppNotify sends a timer event to the application task that schedules it.

From the main loop of the task, we check the event type. If it is a timer event, the application calls ProcessTimers (Listing 6.5).

Listing 6.5   Process the timer event.

```
ProcessTimers ()
{
    Decrement the current timer count in the first entry
    of the timer list;
    If the count is 0 {
        For all the entries in the timer list whose current timer
            count is zero {
            Process timer expiry by calling the timeout routine with
                context provided in the timer block;
        }
    }
}
```

We note that processing the timers in the main loop is preferred to processing the timers in the notification routine. The timeout routine in the timer block could involve a significant amount of processing, including constructing and transmitting packets, processing lists, and calculations. Since it possible that the notification routine can be called from the interrupt context, it is preferable that time-consuming activities be kept out of its execution path.

6.2.3 Timer Management Task

In the previous example, each task maintains its own set of timers by using a set of timer blocks in a timer list. Each of these tasks needs to be notified of a timer tick. This solution uses a large number of context switches to decrement a count in a timer block, since each application to be notified can be a separate task. Also, for each of the applications notified for a tick, we need a separate counter. In the example, AppClock maintains a counter AppxTime for each of the applications requiring a timer. This could be necessary if the granularity of a tick varies. One application may require a one-millisecond tick while another requires a one-second tick. While this is more flexible, it gets complicated if a large number of tasks need timers.

One way to address this is to use a single timer management task. Figure 6.10 illustrates a timer management task, with a tick equal to the lowest of all the ticks required for the various tasks. There is only one change to AppClock, that is, to provide a tick to the timer management task (TMT). The TMT, represented in Figure 6.10, separates the timer tick types by granularity. Assume the need for a 1-millisecond tick, a 10-millisecond tick, and a 1-second tick. The TMT will be notified of a 1-millisecond tick only. It will maintain counters for simulating a 10-millisecond and 1-second timer. For each of these tick types, there are separate timer lists.

Figure 6.10   Timer management task.

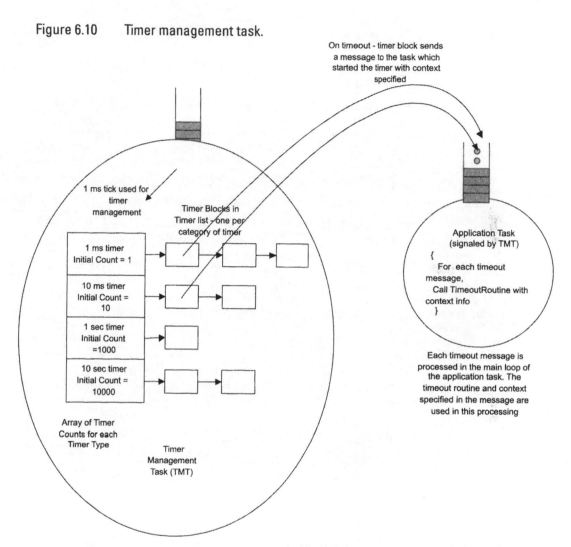

The context in each timer block is now expanded to include the application task which requested the timer. When the timer expires, the TMT sends an event to the specified task in the timer block and passes along information such as the callback routine and parameter list. The TMT sends this as a message to the specified task.

Note that in Figure 6.10, the TMT receives one event, a timer tick expiration for 1 millisecond. The data structures for timer blocks are as specified earlier. An array is used for entries for each of the granularities—1 millisecond, 10 milliseconds, 1 second, and 10 seconds. The initial timer counts for each of the entries are 1, 10, 100, and 1000, respectively. For each tick, each of the timer counts is decremented. When an entry in the array reaches zero, the first timer block for this array entry has its count decremented. If it reaches zero, the timer processing event is called. The logic can be summarized as in Listing 6.6.

Listing 6.6   Create a new timer.

```
AddTimer (TimeoutValue, ApplicationModuleId, Param1, Param2....)
{
    Allocate a timer block and fill in the fields with
        the  task id, and parameters;
    Determine the granularity of the timer from the timeout value
    specified and access the array entry corresponding to the
    granularity (1 ms, 10 ms, 1 sec and 10 secs);
    Add the timer block to the list in the array entry
        using the differential timeout approach;
}


ProcessTick ( )
{

    Decrement the counts in all the array entries;
    If count reaches zero, decrement the count field in the
        first timer block of the list for the array entry;
    If  the decremented timer count reaches zero {
        Delete the block(s) from the timer list and
            indicate the timeout with the parameters
                from the timer block to the task which started the timer;
        Return the timer block to the timer block free list;
    }
}
```

Comparing the Approaches

The first approach, which called for having timer management done in the individual tasks, is quite flexible. A library can be developed that will be used by tasks to allocate and free timer blocks and for maintaining timer lists. The disadvantage is that each task needs to be notified of a timer tick and do its own timer list processing.

The TMT centralizes management but has its own drawbacks. The TMT needs to send events/messages to the applications on timer expiration, requiring additional resources from the system. In addition, the task can itself become a bottleneck if it needs to handle a large number of timers.

Third-party stacks that are licensed and included in the software subsystem usually employ the first approach. All that the new protocol stack requires is a timer tick. The timer processing is handled internally within the stack, usually with a differential timer scheme.

System Issues in Timer Management

The timer management schemes discussed above are quite powerful because of the stored context in the timer block. When the timeout occurs, the application can just use the context from the timer block and perform the timeout action(s).

When the TMT is employed, as shown in Figure 6.10, this task constructs a message with the context provided when the timer was started. It then queues the message to the application task which started the timer and requested notification. The notification is queued to the application task as a message. When the application task processes this message in its main loop, it looks into the context information and determines the time-out routine that should be called.

This flexibility comes at a price. Each of the timer management schemes can allocate and free a large number of timer blocks with each timer block requiring space for application context. This can create a significant memory requirement, especially when there is a large number of timers to be maintained, as with connection-oriented protocols such as TCP. The developer should identify the memory requirements for timers up front and determine the impact on system resources.

If the timer block uses a pointer to reference the context instead of storing it inside the block, it can save some space. This approach is preferred in situations where no context is required—say, when a task just needs to be woken up on a timer event without any additional processing. In this situation, the task does not need to start a timer with a set of parameters or a context.

Checklist for Timer Management

The following can be used as a checklist for the timer management strategy:

1. Use timer management per task if each task has a complex set of requirements for timers—otherwise, use a common timer management task.
2. Choose the minimum number of application timer ticks for the applications, so that you can have a small number of timer lists (see Figure 6.10). Optimize this using the calculated memory requirements for a timer block.
3. Implement a differential timer scheme for each of the timer lists.
4. Connect up to the RTOS timer ISR to obtain the application tick for the timer list.

6.3 Summary

Buffer management is used to allocate, manipulate, and free buffers in the system. The basic premise is to minimize data copying. Local and global buffer pools can be used as well as multiple buffer pools each with its own buffer size. A two-level hierarchy, as in the mbuf scheme, and a three-level hierarchy as in the STREAMS scheme are both popular. An typical buffer management scheme can use control blocks to anchor the configuration, status, and statistics information for the message and data blocks.

Timer Management can be implemented using a system tick from the RTOS. It can be implemented per task or via one single task called the Timer Management Task. The differential timer scheme helps avoid the overhead of linked list traversal in decrementing timer counts. Timer block memory requirements are to be analyzed for efficient system design.

6.4 For Further Study

Keshav (1997) provides a detailed treatment of protocol implementation, with some discussion about avoiding the copying of data. McKusik (1996) provides a comprehensive description of the mbuf scheme. The *AT&T STREAMS Programmer's Guide* is the comprehensive guide to STREAMS.

The TICS Web site provides a tutorial on timer management.

6.5 Exercises

1. Construct a spreadsheet to demonstrate memory requirements for a two- and three-level buffer scheme with all buffers of the same size (256 bytes).
2. Which RTOSes use the mbuf scheme? Are there any variations?
3. Are there any time constraints for timer tick processing? Explain.
4. List the timer features of the RTOS you are using on your current project.

CHAPTER 7

Management Software

Chapter 4 described management interfaces in the context of protocol software interfaces and outlined how the management agent and manager communicate. It also indicated the type of information exchanged. This chapter details the components of the management subsystem for an embedded communications device and its implementation, with a specific focus on device management. Supporting issues such as configuration saving and restoration are also covered.

7.1 Device Management

Figure 4.3 outlined an embedded communications device managed by SNMP from a remote management station. Not surprisingly, SNMP is only one of several protocols and schemes used for device management. Though it is common in the data networking devices, it has not seen wide deployment in telecom equipment. The telecom world uses several protocols, including Common Management Information Protocol (CMIP), Common Object Request Broker Architecture (CORBA), and Transaction Language 1 (TL1) for management of the devices. In addition, a Command Line Interface (CLI) is often used by the operator to configure devices. In the last few years, using a Web browser and HTTP for device management has been gaining popularity. In this method, a network manager connects to the embedded communications device using a browser which presents a screen to configure the various parameters for the device.

7.2 Management Schemes

The large number of management schemes grew out of the history and involvement of multiple standards organizations.

CMIP was specified by the ITU-T for managing networks. It is a machine-to-machine protocol, like SNMP, with a manager–agent paradigm. The Transaction Language 1 (TL1) method of configuration was specified by Bellcore (now Telcordia) as a Man Machine Language (MML) for controlling network elements. It is both a CLI and a protocol, and the PDUs are humanly readable. The most significant reason for its success was that performance reporting and alarms were very well specified. Even though the SONET standard recommends CMIP as the management interface, SONET network elements introduced in the late 1980s had TL1 interfaces. In fact, many of the companies building next-generation SONET equipment in the late 1990s had to support TL1 since that was the only interface the network operators were familiar with.

SNMP was specified by the IETF, which was seen as the key standards body for data networking. Many of the routers and switches first implemented a CLI and an SNMP agent. Though CMIP could have been used instead of SNMP for managing these devices, it never really caught on in the data networking world.

Command Line Interfaces were required for the power users. While some people have likened this to using the DOS prompt when we have the option of a Graphical User Interface (GUI), this mode of configuration survives and continues to thrive. Similar to the TL1 requirement, many of the companies building data networking devices are shipping their products with a familiar CLI interface. The CLI can be accessed via a serial port on the device or via a telnet interface.

Common Object Request Broker Architecture (CORBA) was specified by the Object Management Group (OMG) and is quite popular in enterprise systems. The architecture is similar to a remote procedure invocation between a client and server, with the broker acting as a message passing mechanism. The model was successful because it is programming language independent and uses an object-based approach to data invocation and interchange.

Telecom service providers found CORBA to be a useful way to manage network elements, especially when the elements need to be tied into their own service management and billing systems that are CORBA based. There are some efforts underway to provide translation between TL1 and CORBA for interfacing to legacy systems.

HTTP-based management is often used instead of telnet-based CLI for configuration and control. With HTTP, we have the familiar interface of a Web browser and a less cumbersome way to configure parameters. A pulldown menu can contain only the valid values for a variable, thus removing the possibility of user error, since no other value can be provided as input.

Extensible Markup Language (XML) is used as another scheme for managing communications devices. XML permits a machine independent way of data interchange across machines. XML translator software on the peer systems interpret the XML data according to the context. The model is similar to CORBA-based management but with XML traffic sent over an HTTP connection.

7.3 **Router Management**

Figure 7.1 shows the architecture of a router with SNMP, CLI, and HTTP-based management used to control and configure the device. The term *agent* describes the task or module on the device which terminates the management protocol and makes a request to the protocol or system task. Figure 7.1 shows three separate agents—the SNMP, CLI, and HTTP agents. Each of these agents is a separate task.

Figure 7.1   Router architecture with various management schemes.

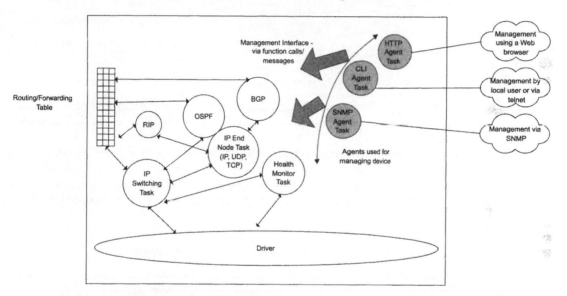

7.3.1 **SNMP Management**

The SNMP agent interfaces to UDP to obtain the SNMP PDUs, verifies the PDUs for correctness and determines the action to be performed. The protocol uses multiple types of PDUs to manage read-only and read–write (settable) variables. Read-only variables can only be queried and cannot be modified by a manager, unlike read–write variables. The SNMP agent enforces this rule in case an errant manager sends a request to modify a read-only variable.

The agent is also responsible for authentication. Earlier versions of the SNMP protocol used a simple community string (similar to a text-based password) to authenticate the manager and enforce access permissions. SNMP Version 3 (SNMPv3) has improved on this to use both user-based and message-based security. The SNMP PDU processing is done at the front end of the SNMP agent. The back end is invoked when the protocol verification and sanity checking are completed. The back end will then determine the required action for the PDU. There are two types of request message from the manager: a "get" and a "set" of any of the variables in a Management Information Base (MIB), discussed in Section 5.1.1. The agent can also send a notification to the manager via a "trap" message.

7.3.2   CLI-Based Management

CLI commands are input by the user and verified by the CLI Task/Agent. The actions to be taken by the CLI task based on user input include:

- Enforce authentication and access rights (user can only display and not modify variables)
- Parse and validate the CLI commands
- Determine the action to be performed
- Make calls to the system/protocol task

7.3.3   HTTP-Based Management

The management mechanism for HTTP is quite similar to the other two agents. Here, an embedded Web server on the target device receives and sends HTML data with management information.

All three methods, while different on the front end, are the same on the back end (as shown by the thick arrows from the three agents in Figure 7.1) They need to make calls to the protocol or system task to perform the operation (reading or modifying variables). This commonality is used to describe the typical architecture of a management subsystem.

7.4 Management Subsystem Architecture

This section details the architecture of the management subsystem in a communications device. SNMP is used to illustrate the architecture and access mechanisms. The principles are also applicable to other types of management, including HTTP, CLI, CORBA, TL1, and so on.

7.4.1   Using SNMP

Chapter 5 discussed how to map MIB-based tables to the implementation. This discussion continues with the example of the IPStatsBlock originally specified in Chapter 5 and detailed again in Listing 7.1.

SNMP provides the ability to get an individual element of a table or individual variable using the Get PDU function. This PDU is verified and translated by the SNMP agent on the device, after which it calls routines provided by the protocol tasks to get the value of the specific variable. These routines are termed the *low-level routines* since they operate at the lowest level in the calling sequence to perform the required function.

So, if the IPStatsBlock is implemented inside the IP task, the task needs to provide a low-level access routine for each member of the IPStatsBlock structure. A Get routine for the number of received packets would look like:

```
IP_Mgmt_Get_ipInReceives ( )
```

Listing 7.1   Statistics block.

```
typedef struct {
            UINT4       ipInReceives;
            UINT4       ipInHdrErrors;
            UINT4       ipInAddrErrors;
            -------
            -------
            UINT4       ipOutDiscards;
            UINT4       ipOutNoRoutes;
            -------
            -------
            -------
            -------
        } IPStatsBlock;
```

For clarity, we simply reuse the MIB variable ipInReceives in the function prototype of the low-level routine—a common practice. Since this management variable is not defined on a per-interface basis, there are no parameters to be specified in the Get routine.

The IP task implements the low-level routines for each of the variables in the IPStatsBlock. The mapping between the MIB variable and the low-level routine to be called is done by the agent through a table.

The second type of routine is a Set routine used for read-write variables. This routine takes at least one parameter, the value for the variable. In cases in which the variables are specified in a table, the routine also requires the indices for the table.

A third type of routine commonly used in the management subsystem is a Test routine. This is invoked before the Set operation by the SNMP agent to determine if the Set command will be successful. It is a good place to validate the parameters.

Consider enabling and disabling a protocol using a variable called IPStatus. Valid values for IPStatus are 1 (enable) and 2 (disable). If a manager requests a PDU to set the value to 3, the Test routine indicates an error, and the PDU returns an error to the manager.

Instead of requiring a routine per variable, we could use a single routine with a switch statement for accessing the appropriate variable. The following example illustrates a switch statement using the fields of the IPStatsBlock (Listing 7.2).

Listing 7.2   A Test routine with a `switch` statement.

```
IP_Mgmt_Stats_GetRoutine (Variable)
{
    ---------
    ---------

    switch (Variable) {
       ----

       ---

       case IP_INHDR_RECEIVES:
               Access Variable and return value;

       case IP_INHDR_RECEIVES:
               Access Variable and return value;

           ----

           ---

       default:
               ------

               ------
    } /* end switch */
}
```

Note that this does not really reduce complexity, since the same logic is needed to access the variable. On the other hand, using a single function can lead to *function bloat*, always an issue with code maintainability. The recommendation is to use a function for each variable. Also, note that a function prototype with the variable name at the end (like `IP_Mgmt_Get_ipInReceives` above) increases the readability of the code.

7.4.2 Using the CLI

The scenario discussed earlier with SNMP is valid for other agent tasks as well. The back end of the agent invokes the same low-level routines, similar to the SNMP agent. The low-level routines are invoked, sometimes in a repetitive manner. Consider the case of a CLI for a router in which the user requests that all the IP statistics be displayed, as follows:

```
CLI> show ip stats
```

The CLI agent will parse this command and interpret that it needs to make multiple calls to the IP task to obtain all statistics variables in the `IPStatsBlock`. It will make multiple calls to the IP task low-level routines—one for each of the variables—and construct a response for the CLI user, as shown in Figure 7.2.

The CLI statistics structure is populated by multiple calls to the protocol (IP) task, after which a display routine is called to display the statistics to the user. The parameters that are to be displayed by the above command can vary depending upon the CLI implementation. For example, the variables to be displayed need not have a one-to-one correspondence with MIB variables. In that case, the variables accessed by the low-level functions need not all belong to the same protocol data structure or MIB table.

Figure 7.2   The CLI agent.

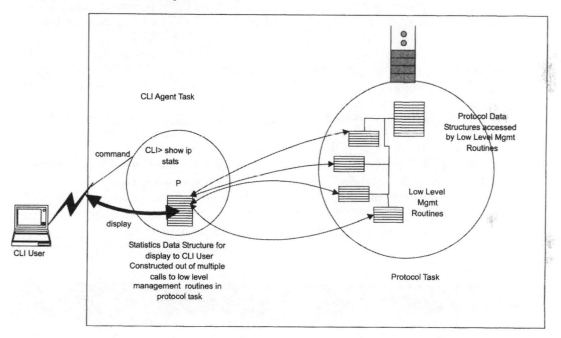

The front end of the agent task is where aggregation of the response from the protocol task takes place. However, access to the protocol task is through a single low-level routine, as shown in Figure 7.2.

7.5 Agent-to-Protocol Interface

Even though the protocol task provides routines for management, these routines are only called from the agent(s), as shown in Figure 7.2. These routines access the various data structures that the protocol task implements. There are two related issues to consider here:

- Priority of the agent task(s)
- Data structure access by the low-level routines conflicting with access from the main loop of the protocol task

The agent task is often assumed to be the lowest priority task since management operations are treated as lower priority activities as compared to the main system functions, e.g., switching or routing. This implies that the low-level routine of a protocol task often executes at a priority lower than the protocol task itself. This can be a problem if the same data structures are being accessed. Mutual exclusion via semaphores is one way to address this problem. The semaphores can be used by the low-level routines and the protocol task routines for access to the data structures.

Semaphores would require significant modifications to the protocol code for management access to be "safe." Another option is to lock the agent task during modifications so that it is not preempted by the protocol task. This ensures that the management task completes its low-level routine access before the protocol task is scheduled, ensuring that the data structures are not corrupted. This can also be realized by raising the priority of the agent task so that is higher than any of the protocol tasks during the time we need to access the read-write variable. After the access is completed, the code can make a system call to return to the original priority of the management agent task.

The ideal solution is to have the management access happen in the main loop of the task. This is similar to the method of using a call to send a message to the task, as in the case of messages and timers, discussed earlier. Lateral access to the data structures is prevented since the management message is processed in the main loop. Since there will be only one thread of execution, there is no need for mutual exclusion.

To implement the management functions in the main loop, we can use the following procedure:

1. Call the low-level routine in the protocol task from the Agent Task
2. Low-level routine sends an event to the protocol task after identifying the type of operation required.

Legacy Systems

While the approach of using a single low-level access mechanism may appear intuitive, it was not always implemented this way. Several systems (some still deployed) had two paths for management—one for SNMP and one for CLI. SNMP support was used only for some of the standard MIBs (like MIB-II), while the "real" configuration was done via the CLI.

Before we blame the developers, we should understand that they often did not have a choice. Some of the standard MIBs at the time (like MIB-II) did not support sub-interfaces like PVCs on an interface. Some of the MIBs also had the problem of not making some variables read–write as required by the network operator. Rather than implementing a proprietary version of the standard MIB, designers simply sidestepped the issue and provided flexible configuration only via the CLI. Often times, designers have to work with constraints along with legacy/history, and their choices may not always be optimal.

In Figure 7.3, the low-level routine is called from the SNMP agent after decoding. Instead of the low-level routine accessing the data structures of the protocol task, the routine constructs a message with the type of operation requested and parameters for the operation and then sends the message to the protocol task queue. This is a *local event* or *internal event* on a specific task queue, which enables the task to handle it in its main loop. Once this message is queued, the protocol task is scheduled, since it has a higher priority than the SNMP task and is waiting on events on its queues. The management function is now handled in the protocol task main loop.

Figure 7.3   Management Routines and Internal Events.

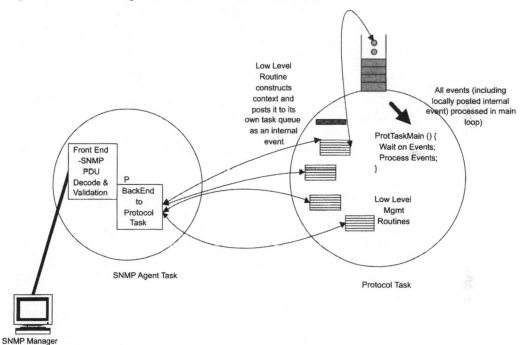

The value can be returned to the agent task with another message. For this, the protocol task constructs a message and sends it to the agent task with such information as the name of the calling routine (which initiated the internal event), status of the operation, and any return values. The agent processes this response in its main loop, since it is a message posted to one of its queues and is required for a pending manager query. The return code and values from this message are used to construct a response to the manager (SNMP Manager or CLI) which initiated the request.

7.5.1 Memory Separation Between Agent and Protocol

If the agent task and the protocol tasks are in two separate memory spaces, we have to construct a message from the agent task and queue it to the protocol task. We cannot call the low-level routines of the protocol task from the agent tasks since they are in separate locations. This procedure is no different from other internal events, except that the agent task explicitly constructs the message.

The response from the protocol task to the agent task is the same as before. The protocol task constructs a message in response to the query and queues it to the agent task. The agent dequeues the message and maps it to the pending manager query and sends a response to the manager.

This effectively implies that we use a message-based interface instead of a procedure-based interface. This approach scales very well to multi-board communications systems. The agent and the protocol task may be on two separate boards, which is an extension of the separate memory space idea. Multi-board communications systems and their design are covered in Chapter 8.

7.6 Device-to-Manager Communication

In addition to communication initiated by the manager or operator via SNMP, CLI, or HTTP, agent-to-manager communication is another important part of the management model. For example, a manager may be alerted when a port goes down or when traffic on a port exceeds a threshold. This unsolicited alert is sent on the same communication channel as the manager-to-agent interaction. SNMP uses a mechanism called a *trap*. With a CLI, the alert is printed on the user console.

The type of information the operator requires, the filters on the alerts, action taken on the alerts, and so forth are all areas that merit additional discussion. However, the significant issues related to an embedded communications device are:

1. The protocol task uses the same queuing mechanism for traps or alerts as it does for get or set responses. The protocol task does not send traps directly to the manager. Rather, it sends alerts to each of the agents (CLI, SNMP, CORBA, etc.). The agents will determine when to pass along the alert.

2. Agent design should restrict the number of traps to space them such that they are more manageable. One technique is to avoid sending a trap to a manager if you had sent it the same trap, say, 5 seconds ago. This is known as throttling or rate limiting

3. The alerts/traps should be grouped into *critical* and *non-critical* at the protocol or system task level. The design should ensure that critical alerts are sent before thresholds are exceeded rather than after depletion. This allows the manager to gracefully shut down the system.

This device may be one of several systems which are sending alerts, so be sure to use alerts and traps sparingly and only for critical events

7.7 System Setup and Configuration

The previous discussions focused on the design of the management subsystem for an embedded communications device. This section details configuration and management issues related to system startup.

7.7.1   Boot Parameter Configuration

Chapter 2 discussed using flash images to download an upgrade from a remote host. The system needs the boot monitor to be configured with the IP addresses of the device host to download the image, user name and password for FTP access, interface to use for the boot, and other boot flags as listed below. In a VxWorks™ system, this is a stripped down version of the OS, often termed the `bootrom`, from the name of the image in the `make` file.

Typical boot parameters for a VxWorks™ boot are shown below, although this is not a complete list:

- Ethernet IP Address
- Boot File Name
- Remote Host IP Address
- Gateway (Router) IP Address
- FTP User Name
- FTP Password
- Startup Script
- Boot Flags

Additional parameters can be included by creating a new version of the boot ROM and modifying the processing logic. The user needs to configure parameters using a local (serial) terminal and then initiate the boot. The download is performed using a protocol such as FTP or TFTP, implemented in the bootrom. Boot parameter configuration is done using a simple command-line interface, which may have no resemblance to the final CLI used in the device.

It is not possible to use SNMP, HTTP, or any other interface to perform this configuration, as the system does not yet have an IP address to use for these types of communication. When the image is downloaded to flash, the boot parameters can be changed so that subsequent boots are performed from the image in flash, without having to reinitiate a download.

7.7.2   Post-Boot Configuration

Once the image is downloaded, boot parameters can still be changed using SNMP or other mechanisms. After a change, the user saves the changed parameters to EEPROM or flash using SNMP or a CLI. This helps avoid the need for local configuration on the next boot. When the system reboots, it picks up the new boot parameters and downloads the appropriate image.

The management subsystem can also provide a variable which can be set by a manager to initiate a system restart. This can be done for multiple reasons: a misbehaving process, use of a new image or new configuration parameters, restore to default configuration, and so on.

7.8 Saving and Restoring the Configuration

It is important to be able to save the existing configuration for an embedded communications device and to restore it on startup. Protocol configuration is quite complex, but saving the current configuration and restoring it upon startup makes the process easier. Save and restore processes can be either manager or device based.

In a manager-based approach, the SNMP manager uses a get operation to keep a record of the current device configuration. When the system reboots, it requests a configuration restoration from its manager. The manager performs this with multiple set operations, effectively "replaying" the saved configuration.

This approach provides a centralized approach to management in which the device is not required to save and restore configuration by itself. The manager has to ensure that it sets the parameters in the right order—basic parameters first, followed by non-basic parameters.

Device-based save and restore operations are more popular. Here, the manager instructs the device to save the current configuration during system execution. The device can categorize and aggregate the basic and non-basic parameters and store them efficiently, as discussed next.

Theory of Operation

When implementing a protocol, ensure that configuration parameters and the method for saving and restoring them are identified up front. Moreover, only the parameters that can be set (i.e., configuration parameters) need to be saved. Variables/parameters related to status and statistics will not fall in this category.

The simplest approach to the saving/restoring of configuration is to operate on the actual data structures. This implies that we can save current data structures and restore them upon startup.

One issue with the approach of saving and restoring data structures is that basic parameters may have to be grouped and set before the other parameters can be set. Data structures often follow a more integrated design, making it difficult to separate the basic parameters out.

Another problem with data structure restoration is that configuration should cause some effects and side effects. In the case of enabling a protocol, the Enable' operation will be much more than just changing the value of a variable to "Enabled." The action can cause PDUs to be sent on multiple interfaces and also effect the starting of various timers. So, it is essential that, when restoring configuration, we mimic the effects of the steps that caused the configuration—effectively recreating the sequence.

Saving the Configuration

The read-write basic and non-basic parameters are aggregated and saved to a temporary location in the RAM, as shown in Figure 7.4. A common method is Type Length Value (TLV) encoding, which specifies the type of the data being saved, the length of the stored value, and the actual value. Prior to storing the file, the binary file is typically com-

Figure 7.4   Saving the configuration.

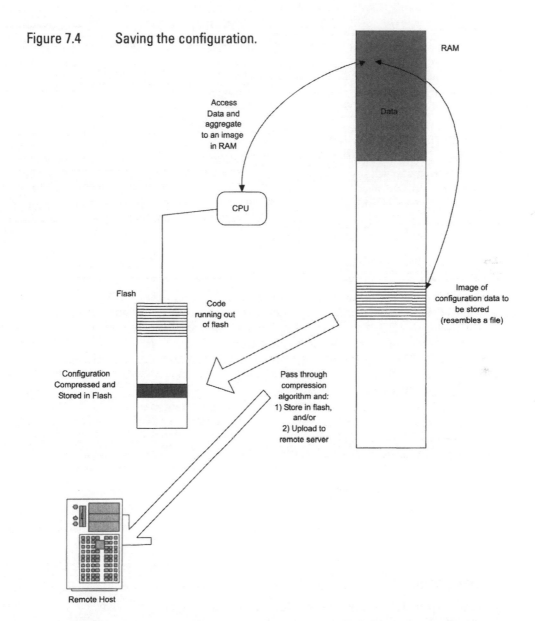

pressed. For error detection, a checksum is calculated on the compressed file and appended to the end of the file. The entire chunk of data, as shown in Figure 7.4, is a binary file.

The file can be stored locally or at a remote location. The upload to a remote location is done using upload parameters like the remote host IP address, file name, FTP user name and password, and so on, similar to the boot image. Good design paradigms provide for both local and remote storage options, since it is likely that the configuration file may be too big to fit into local storage.

Note that when a save is initiated, the system manager will be permitted only to display variables and not to modify them. This is to ensure the integrity of the saved configuration.

Restoring the Configuration

Restoration of the configuration happens at boot time. After the image is downloaded and the system started, the configuration file is loaded and decompressed from the local storage or a remote host. The structures will be used to configure the basic and non-basic parameters for individual protocols. Note that the system is not fully initialized until this process is complete. No configuration or management operations are permitted until the system has completed this operation.

7.9 Summary

Multiple management schemes exist for managing a communication device. SNMP, CLI, and HTTP based management are often used in switching/routing devices. An agent (SNMP, CLI, or HTTP) acts as the front end for the management operations by decoding the PDU or user command and translating it into a call to the protocol task. The call is to a low-level access routine in the protocol, which does a Get, Set, or Test of a variable. The agent-to-protocol interface can take place via procedure calls which can translate into internal events and messages.

Saving and restoring configuration is a very important part of a communications device. The configuration parameters are usually written to a specific area of RAM from where the image is compressed and stored locally or at a remote host. Restoration is the same sequence in the reverse order.

7.10 For Further Study

Stallings (1998) provides a comprehensive description of the SNMP protocol. Cisco Systems' document on saving and restoring configuration on their WAN switches provides a good system vendor perspective on the topic.

7.11 Exercises

1. Enumerate the basic parameters and non-basic parameters for a Layer 2 switch with IP end node capabilities.
2. Elaborate on the performance issues related to a message-based exchange between the agent and protocol tasks.
3. Suggest a TLV encoding for storing an IP routing table in a local file.
4. What are the recommended recovery actions when the restoration of configuration fails?

CHAPTER 8

Multi-Board Communications Software Design

Previous chapters covered issues related to communications software design on single-processor and single-board architectures. However, communications systems can be complex, often involving several boards in a chassis and, in some cases, multiple chassis. The software design will need to handle the distributed architecture of a multi-board system. In most development situations, engineers will not be able to maintain multiple baselines for their software—one for the single-board system and another for the multi-board environment. In these cases, it is useful to design the software to be able to work easily in both environments.

This chapter details some common multi-board designs used in communications and details how the software architecture will change for designs. It will also detail the issue of high availability and redundancy as it is commonly used in multi-board systems. As before, the Layer 2/3 switch is used as the example platform for the discussions.

8.1 Common Architectures for Communications Equipment

Communications devices can differ significantly in their form factors—for example, a cell phone is several times smaller than the telecommunications switch through which its calls may be switched. The following is a partial list of the popular forms for communications devices, along with some examples:

- Host-based adapter—a PCI Ethernet add-on card
- A handheld device—a cell phone or pager

- A "pizza box" design—a single processor or multi-processor switch or router
- A chassis-based design—a CompactPCI-based voice or data switch

Of these, the pizza box and the chassis designs are the most popular for communications devices in a network. The pizza box design is typically a single-board architecture with one or more processors. The chassis-based design has multiple boards plugged into a backplane which provides the communication among the boards.

8.1.1 Single-Board Designs

A *single-board/single-processor architecture* is the simplest design. Here, both the control and data planes run on the only processor or CPU. In several cases, data plane acceleration could be provided by a hardware device, like a switching chipset (an ASIC). In these cases, hardware acceleration devices do not run software—they are only configured by the software running on the central processor. An exception is the network processor, which typically runs microcode downloaded from the central processor.

The second type of architecture is the *single-board/multi-processor architecture*. The simplest example is the two-processor architecture, in which the control and management code run on Processor 1, and the data plane and services run on Processor 2, all on the same board. This separation isolates the core function of the system (data plane) from control and management functions.

Consider the case of a Layer 3 switch (IPS) with two processors. Processor 1 runs RIP, OSPF, the SNMP Agent, and the CLI Agent. Processor 2 runs the forwarding function services, like NAT and Firewall, with support from the hardware. Packets will first be processed in Processor 2 and dispatched to Processor 1 if they are targeted towards the control plane or management plane (as represented by the routing protocol tasks and management agents). Processor 2 will communicate with Processor 1 using Inter-Processor Communication very similar to communication in a single-processor system. Also, a segment of memory can be shared between these two processors and used for data exchange between them.

The two-processor example can be extended to include multiple processors, all on the same board. Or, the board can possess slots to permit add-on cards. Each of these add-on boards can have a processor and use proprietary mechanisms to communicate with the base board. The PCI Mezzanine Card (PMC) is a variation on this concept. The PMC is added parallel to the main board in a PMC slot.

8.1.2 Chassis-Based Designs

The most general form of multi-processor architecture is the use of multiple boards in a single chassis. An example of a popular architecture in communications is CompactPCI, specified by the PCI Industrial Computers Manufacturing Group (PICMG). The first version of this specification used the PCI electrical specifications in an industrial form factor and the PCI bus for inter-card communication. Follow-on specifications like PICMG 2.16 included a packet switching architecture for inter-card communications.

Variations of this design include addition of a H.110 Time Division Multiplexing (TDM) or PICMG 2.15 bus for carrying circuit-switched traffic. So, the same multi-service design (data + voice) may have a packet-switched interconnect for data and a H.110 bus for carrying TDM traffic. A design with a packet-switched backplane and H.110 bus is shown in Figure 8.1.

Figure 8.1 Chassis design with packet + circuit switched buses.

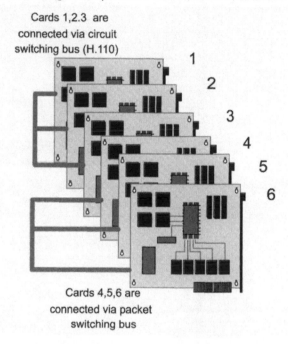

The isolation between the TDM and packet traffic is required because TDM traffic requires a guaranteed delay, which is achieved by using fixed slots on the H.110 bus for passing the voice traffic between the transport cards. Similar to the software model of control and data traffic between tasks, multi-board systems typically use different transport mechanisms for control and data (or payload) traffic. In a chassis based router, it is quite common for the control traffic to use a different transport than the payload. The PCI bus is one example of a control interconnect. Fast Ethernet (100 Mbps) or Gigabit Ethernet (1 Gbps) are also used in some systems for control traffic within the chassis.

8.1.3 Rack-Based Designs

Variations on the pizza box– and chassis-based designs find application in rack environments. It is quite common to find router or switch devices with a 1U or 2U form factor mounted in a *rack* (see Figure 8.2).

Figure 8.2   A multiple-card rack design.

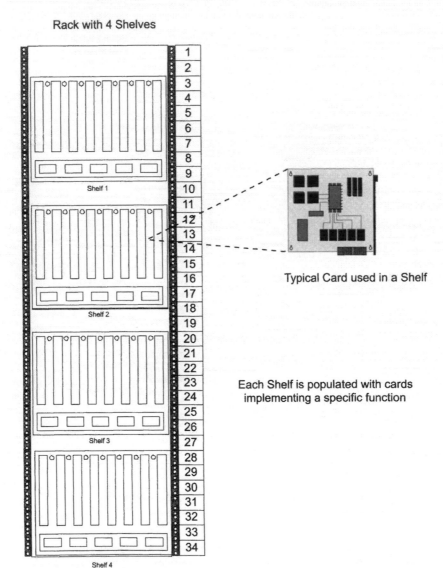

Rack with 4 Shelves

Typical Card used in a Shelf

Each Shelf is populated with cards
implementing a specific function

A rack is a standard cabinet used for mounting servers, network gear, storage, or other components in a complete solution. Hardware vendors use the standard sizes when building any hardware that is intended to be rack mounted. Racks are designed as empty cabinets with rails for mounting equipment, and often racks are designed to handle the electrical and thermal requirements of specific applications by including power supplies and cooling fans inside the cabinet.

Each rack can house multiple devices—the height of which is expressed as a U form factor, or 1.75 inches high for each U. Cabinets (racks) can contain 42U or more, depending on the application and environmental requirements. A high-density server may contain several CPUs but take up only 1U (1.75 inches) in a cabinet. CompactPCI cards are often positioned vertically and require several Us, depending on the architecture. For example, a common configuration for a large system is a rack with multiple shelves, each shelf housing multiple cards, as shown in Figure 8.2. Each shelf can be considered to be a single system or a subsystem, depending upon the complexity of the communications device. A computing or communications solution can also occupy more than one rack unit, as shown in Figure 8.3.

The connections between the individual cards can be through a backplane or a midplane architecture. The backplane architecture indicates that each card is of the full 15.5 inches in depth, while the midplane architecture implies that each card has a depth of 7.75 inches. The system can be expanded by using fiber optic cable between the individual racks. Protocols, several of them proprietary, have been developed for the inter-card communication.

8.2 Multi-Board Architectures

Multi-board architectures are needed since a single-processor/single-board system cannot scale to handle a large number of ports or increasing port speeds. While hardware acceleration could be used to address the port speed problem, there is a limit to the amount of board real estate available for high port densities.

Modularity is another reason for multi-board architectures. The customer needs to be able to add or remove ports from the system as his/her requirements change. In this scenario, an enterprise or service provider can purchase a low-end multi-board system with only a few I/O slots populated. Based on his need, the customer can add more cards to the system to expand the system's capacity and capability without having to throw out the original investment. This is the *"pay as you go"* philosophy which many equipment vendors use. There are several variations of the multi-board architecture. This chapter focuses on the functional organization and interconnects using a Layer 3 switch/router, also termed the IP Switch (IPS) to illustrate the concepts.

8.2.1   Components of a Multi-Board System

The most common types of cards in a multi-board communications system are the *control cards*, *switch fabric cards* and *transport cards* (often called *line cards*). Control cards have the following functions:

- Running control (e.g., routing protocols such as OSPF and BGP) and management protocols. The messages generated by these protocols are sent out on line card ports.

- Building tables using information from control and management protocols for use by the line cards. The tables are provided to the line cards using card-to-card communication.
- Providing an interface to management through tasks such SNMP, CLI, and TL1 Agent.
- Performing system management functions such as shelf and chassis management and internal card-to-card communication monitoring. These are required to keep the system running.
- Managing redundancy when switching to a new line card.

Line cards are responsible for:

- Sending and receiving messages on the hardware ports.
- Switching and forwarding traffic between line card ports using the tables built up by the control card.
- Sending the received control and management messages to the control card.
- Providing local control and management information to the control card, e.g., indication of a port failure.
- Managing redundancy when switching to a new control card.

Most systems are designed to be modular by providing the media interface as a separate plugin module to the line card instead of integrating it on to the line card. This permits upgrade of the interfaces without having to upgrade the line card. For example, an 8-port Fast Ethernet module can be replaced by a single Gigabit Ethernet module without having to replace the line card (see Figure 8.4). This also implies that the module-to-line card interface is compatible with and able to handle the higher speed Gigabit Ethernet interface. For this chapter, the term *line card* indicates the line card plus its media module.

This functional separation of control card and line cards is applicable to several types of communications equipment, though the actual systems may use different terminology to describe them. Another way of looking at this is to visualize a split between the components of a single CPU system. *The "closer to the wire" components are now housed in the line card, while the "closer to the application" components are housed on the control card.*

The control card and line card correspond to the functional separation of control plane and data plane, as described in Chapter 3. This discussion assumes that part of management plane functionality is now included in the control card.

The third type of card is the switch fabric card. This is a card housing the functionality to switch traffic between different line cards. This card avoids the need for a "full-mesh" relationship between the line cards, in which a line card needs to be connected to every other line card. Instead, the line cards are now connected to the switch fabric. The switch fabric typically operates on fixed-size cells, so a line card will slice its packets into these fixed-size cells before sending them to the fabric.

There are two common architectures for multi-board IPS systems using the control card and the line cards:

1. Single Control Card + Multiple Line Card Architecture
2. Multiple Line Card Distributed Architecture

The following sections discuss these architectures in greater detail.

8.3 Single Control Card + Multiple Line Card Architecture

An example of the single control card + multiple line card architecture is shown in Figure 8.4. There is one control card, with four line cards and a switch fabric card. The switch fabric and the line cards are connected through high-speed serial links. The control card and the line card can be connected by PCI or Fast Ethernet. The latter is sometimes known as the internal Ethernet or *Ethernet on the backplane*. The control card–to–line card interface is used for messages that need to be processed by the control card's processor(s) and for messages originated by the control card to be sent to the physical ports. The bandwidth requirement on this control card–to–line card interface is not as high as the line card–to–switch fabric interface. However, when systems with a large number of line cards are built, the traffic on this interface can become an issue. System designers often address this by using a high-speed switched connection like Switched Gigabit Ethernet or Infiniband.

8.3.1 Line Card–Line Card Communication

Line card–to–line card data transfer is through the switch fabric. The switch fabric card provides the switching function for the payload sent by the line cards. In the IPS, if a frame is forwarded from Port 1 of Line Card 1 to Port 3 of Line Card 2, it is sent on the serial link from Line Card 1 to the Switch Card and then from the Switch Card to Line Card 2. A header on the frame indicates the port number on Line Card 2 where the frame is to be sent. The line card (switch fabric electronics) usually has two parts—a line card component and a switch fabric component. Second, the frames are sliced into fixed-length cells and sent on the link. Switching is done faster with fixed-size cells, and the delay is predictable.

8.3.2 Line Card–Control Card Communication

A messaging interface is used for interactions between the tasks on the control card CPU and the tasks on the line card CPU, as shown in Figure 8.3. The types of interactions in Table 8.1 take place on this messaging interface.

Table 8.1   Messaging interactions.

Type of Message/Function	Direction of Communication	Examples/Comments
Message generated by Control Card for transmission on external port	Control Card to Line Card	Routing Protocol Updates, Keep Alive
Control/Management message received on external port	Line Card to Control Card	Routing Protocol or Spanning Tree Protocol messages
Control/Status Message Request	Control Card to Line Card	Query for statistics information on a specific port
Control/Status Message or Response	Line Card to Control Card	Message indicating status change of a port, response for a statistics query from Control Card
Payload message after CPU processing	Control Card to Line Card	MAC frame forwarded on all ports if destination address is unknown
Payload message requiring CPU processing	Line Card to Control Card	MAC frame with Unknown destination address
Heartbeat Message	Control Card to Line Card	Used by Control Card to test whether the line card is functional and also for testing the link between the two. Used for redundancy
Heartbeat Response Message	Line Card to Control Card	Sent by line card to indicate the current status

As the table indicates, each message type from the control card to the line card has an equivalent message type in the reverse direction. A heartbeat message is typically sent as a request from the control card to the line cards. The line card has to respond to this heartbeat request with a response in the reverse direction. If the control card has not received responses to a series of heartbeat messages, it can timeout the line card and flag it as down—note that the control card can only report this as an error but not determine if the line card has failed or if the link has failed.

Protocol messages destined for tasks on the control card are forwarded by the line card. In the IPS, these could be control plane packets like routing protocol PDUs and packets destined to the IPS, such as TCP/telnet packets The line card will make the distinction by comparing the destination IP address in the packet with the IP address of the switch ports or the broadcast/valid multicast address. If there is a match, the packet will be passed on to the CPU (i.e., control card)—the typical IP end node operation.

Figure 8.3   Single Control Card + Multiple Line Card Architecture.

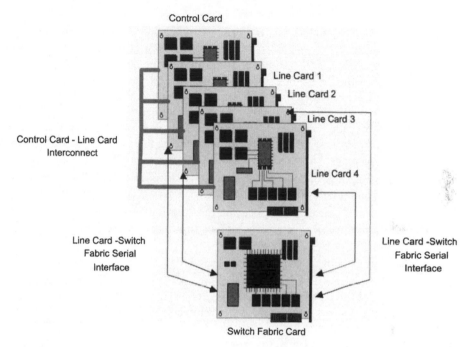

8.3.3 Message Types and Protocols

Table 8.1 indicates a simple summary of message types required on the interface. In more complex scenarios, the control card can affect the forwarding behavior of the line card. This can require additional types of messages from the control card. One scenario is the modification of the traffic management parameters on the line card. This can be the changing of the queue sizes and/or algorithms for packet discard. This area is being investigated by the Forwarding and Control Element Separation (ForCES) Working Group of the IETF. ForCES is specifying the protocol and the format and semantics of the messages between the control element and the forwarding element in a communications system with plane separation.

The message protocol used between the control card and the line card is usually proprietary since it is only used "inside the box." However, some systems do run protocols such as TCP to provide reliable communication between the control card and the line card, but this is inefficient since TCP is a general-purpose protocol.

While TCP (or a proprietary protocol) may be used for transport, there is a need to have another protocol for card-to-card communication, Inter-Card Communications Protocol (ICCP), which will run over the reliable transport protocol, like TCP. This protocol will identify the type of message, specify the command or response, contain parameters related to the messages, and so on.

8.3.4 Partitioning Software Between Control Card and Line Card

Chapter 3 detailed the functional architecture of a Layer 3 switch or router. This was a single-processor design in which tasks interfaced with each other using Inter-Process Communication (IPC), and the driver programmed the hardware controller as well as received and transmitted frames from the Ethernet links (see Figure 3.5). In this section, this basic design is expanded into a distributed architecture using a control card with its control processor and a line card with multiple Ethernet ports.

In Figure 8.3 (page 139), the routing tasks such as OSPF, BGP and SNMP are now located on the control card. These tasks build routing tables and provide information to a Route Table Manager (RTM). The RTM has the following functions:

- It is responsible for *route leaking or route redistribution*, based on a configured policy—e.g., which BGP routes are to be known to OSPF.
- It uses the "raw" routing table/Routing Information Base (RIB) information from the various routing protocols and constructs a forwarding table/Forwarding Information Base (FIB).
- It distributes the FIB to the individual line cards.

The RTM `is responsible for providing the FIB to the individual line cards. It is also responsible for FIB modifications. Moving from the single-processor architecture of Figure 3.5 to the control card + line card architecture of Figure 8.3, we note that the IP Switching Task in Figure 3.5 is now on the line card. This task uses the RTM-provided FIB information to program the hardware on the line card for forwarding, including building the hardware forwarding tables. The details vary with the type of hardware-forwarding or -switching silicon devices.

8.3.5 Partitioning Abstraction

Abstraction implies that the higher layer software does not know task distribution across multiple boards and treats the software as if it were running on a single CPU system. Higher layer software tasks continue to use IPC to communicate with another task, except that the task may be running on another board. The communication mapping to another board is done in the IPC implementation and does not affect the two communicating tasks.

As shown in Figure 8.4, the routing protocol tasks perform as if they are on a single board system with 8 ports. This function is provided in the abstraction layer, which will be the only component changing when moving from a single-board to a multi-board system. In Figure 8.4, we call this a Distribution Abstraction Layer (DAL). Note that the DAL on the control card has a corresponding DAL on each of the line cards.

The DAL needs to perform the following functions:

- Maintain mapping of interfaces to board number + port number

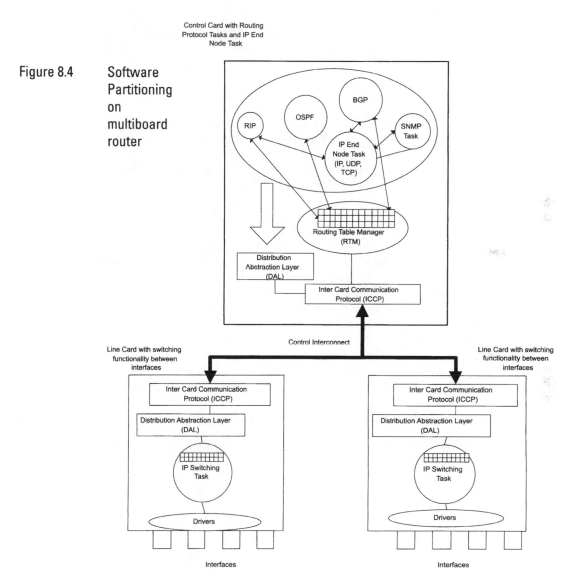

Figure 8.4 Software Partitioning on multiboard router

- Maintain mapping of task and resource names so that distribution is hidden from the higher layer tasks. Application tasks only use the name for the resource—mapping is done at the DAL.
- Provide IPC services across the cards using system services as appropriate

Mapping Interfaces

Mapping interfaces to the board and port number are required by all tasks using the interfaces, both logical and physical. An example of a logical interface ID to the physical port numbers is provided in Table 8.2.

Table 8.2   Logical and physical interface mappings.

Logical Interface ID	Shelf Number	Board Number	Port Number
1	1	1	1
2	1	2	2
3	2	2	1
4	2	2	3

The protocol and system tasks use only the logical interface ID. The DAL performs the mapping so that messaging can be done appropriately. For example, if the RIP task sends out an update over Logical Interface 2, the DAL will map it to Shelf Number 1, Board 2 and Port 2. This mapped information will be used to send the routing update over the control card–to–line card link to Board 2 of Shelf 1.

Mapping Names

A related and more general version of interface mapping is achieved by the Name Database. This is a single repository for the location of all the resources in the system. Consider an IPS implementation in which the control card runs a TCP/IP end node implementation while a line card may run the PPP protocol as a task. To communicate with the PPP task, the IP task on the control card makes a call such as the following:

```
SendMsg ("PPP", msgPtr)
```

The SendMsg routine uses the DAL to determine the location of the PPP task. It does this by using the mapping table maintained by the Name Database, which resembles Table 8.3. The Name string is used to look up the details of the task or resource and determine its location within the system. In the example, the PPP task is on Line Card 1, so the DAL routes the message through the ICCP to Line Card 1. The application calling SendMsg will not know about this translation and routing.

Table 8.3   Interface mapping by the Name Database.

Resource Name	Resource Type	Location (Board/CPU)	Access Parameters/Comments
"RIP"	Task	Control Card	Only one control card in the system running routing protocols
"PPP"	Task	Line Card 1	Only Line Card 1 has serial ports running PPP
"SNMP"	Task	Control Card	Control Card also runs management functions
"IP Switching Task"	Task	All Line Cards	Use Interface ID for distribution to correct line card

When messages are to be sent from the control card out on specific ports on the line card, the DAL uses the logical interface mapping to locate the board and port number on the board to receive the message. The DAL and ICCP may use the services of the real-time operating system for multi-board communication.

8.4 RTOS Support for Distribution

Other than mapping the interface ID to the physical ports, the name database and intercard messaging functions are quite general. Some RTOSes offer these as part of their distribution. For example, the OSE™ RTOS provides a facility called the Name Server, which tracks the physical paths to system services. This allows OSE applications to use logical names for communicating with any service. If there is a change in the distributed system topology, the Name Server is updated. So, applications will not notice any disruption, as they will always use the name for communicating with the services.

OSE also offers a facility called the Link Handler, which can be used to implement the DAL messaging function. The Link Handler keeps track of all resources within a dynamic distributed system. If a process fails, the OSE Link Handler notifies all processes that were communicating with the failed process. The communicating processes can take action on this notification. The Link Handler also notifies the Name Server of the availability or non-availability of resources.

8.5 Data Structure Changes for Distribution

Recall the discussion about accessing global variables via access routines instead of pointers. This approach allows controlled access, in which mutual exclusion can be enforced.

When moving to a distributed environment, there are two options for accessing the variables:

1. Maintain the access routine paradigm. Individual tasks will continue to call the access routines, but the routines will make a remote procedure call to obtain the information, assuming that the data structure is on a different card.

2. Have a local copy of the data structure, and replicate the access routines on both cards. This requires synchronization of the two copies of the data structure.

Figure 8.5(a) details this access routine scenario. The data structure is stored on the control card, and the access routine is implemented in a separate task, so that all access to the data structure is through messages. For both control and line card tasks, the messaging infrastructure takes care of the abstraction.

Figure 8.5(b) details the local copy scenario. The data structures are replicated in both cards, and the access is completely local. This avoids the overhead of messaging across the backplane for basic access to the data structures. However, the data structures need to be synchronized; for every change in one data structure, there needs to be a corresponding change in the other data structure. This can be difficult to maintain when there is a large number of cards which replicate the data structure information.

Figure 8.5   Access routine and local copy scenarios.

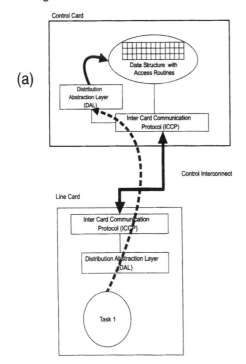 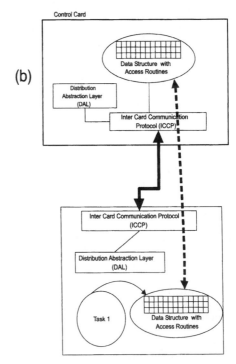

8.5.1 IPS Approach

In the IP Switch (IPS), note that the RTM–to–line card interaction falls in the replication category. Each line card has a copy of the FIB provided by the RTM, which keeps them synchronized. Early implementations of multi-board routers used a subset of the complete FIB on their line cards. The FIBs had only information about forwarding between the ports on the same line card. For all other packets, the line card required a lookup of the FIB maintained by the control card. The pseudocode in Listing 8.1 details the logic.

Listing 8.1   The IPS approach to accessing variables.

```
Look up local copy of FIB
If packet can be forwarded between ports on this card, perform forwarding;
Else
    Look up the destination card for the packet from the RTM FIB on the
        Control Card;
Send the packet to the appropriate line card via the switch fabric
```

This approach allowed the FIBs on the line cards to be much smaller in size as compared to the FIB on the RTM. However, overall performance suffered, since packets destined for other cards had to perform a remote lookup function call to get destination information. Recent implementations keep the complete FIB on the line card, so the forwarding can be done without any lookups to the control card.

8.6 State Machine Changes for Distribution

When moving to distributed or multi-board systems, there are two approaches to the protocol state machine implementation. The state machine can be completely located in one card (*monolithic control plane*), or it can be implemented in two parts—one part on the control card and the other on the line card. The latter approach is sometimes called a *split control plane*.

8.6.1 Monolithic Control Plane

In the monolithic control plane scenario, action routines are called from the state machine implementation, as before. The difference now is that the routines could result in a message sent to the line card. The action routine can wait on a response before returning, i.e., a hard wait is performed for the response while in the action routine. However, the routine cannot wait forever and may need to be signaled on timeout. This is similar to the select call with timeout that is present in BSD UNIX. If the call times out, the action routine returns a failure code. Subsequent error handling is similar to the single-CPU situation.

8.6.2   Split Control Plane

The split-control plane permits parts of the control plane to execute on a line card. For example, the "Hello" processing in OSPF can be implemented on the line card (called the OSPF Hello Task in Figure 8.6) This processing requires periodic polling of neighbor routers as well as processing their polling (Hello) messages. This approach has the following advantages:

- A faster response time to the neighbor, since the "Hello" need not be sent to the control card for processing.
- Lowering the bandwidth consumption on the control card/line card link—since the "Hello" messages are not sent on this link.

The split control plane requires the state machine at the protocol level to change. In the OSPF example, the neighbor state machine is used as the engine for the neighbor PDU processing. Implementations will ideally keep this state machine and table independent of the main state machine table, thus permitting it to be easily moved to a distributed environment. In a split control plane environment, the neighbor state machine will now be implemented in the line card. Inputs to the neighbor state machine will be the same as before except that the action routines may need to send messages to the main state machine on the control card. Similarly, action routines from the main state machine may need to send messages that impact the neighbor state machine on the line card.

Figure 8.6   OSPF Split Control Plane Example.

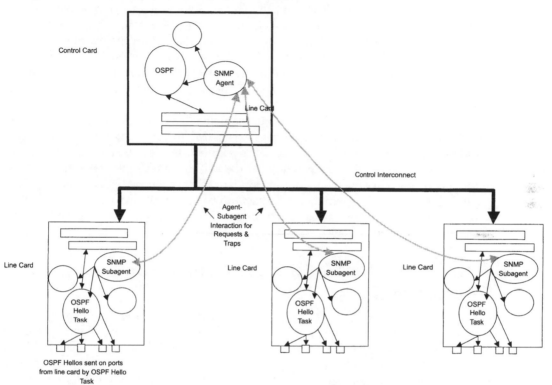

8.7 **Management Interfaces for Multi-Board Software**

Chapter 7 detailed the use of multiple types of management agents. This section covers how the management subsystem architecture needs to be modified for multi-board systems.

The underlying premise for the design is that management agents are always located at a central card. Some systems use a separate card for management function, but, for this discussion, we assume that the management function resides on the control card. The organization is similar to Figures 7.2 and 7.3, which show examples of the agents calling low-level routines. Since several of the managed entities could be on the line cards, the agents need to interact with an equivalent management function on the line card for configuration and control.

This management entity is often called a *subagent*. There is only one subagent for each line card. The management agents on the control card and the subagent on each line card communicate with each other for management operations. To implement this function, the management agent–to–lower layer interface of Figure 7.2 explodes into a multi-board messaging interface. The common agent on the control card will use the messaging capabilities of the ICCP and communicate with the subagent on the line card.

The subagent acts like an agent for the tasks on the line card. This is effectively a DAL extension since the line card tasks have no visibility into the distribution of the management function. Traps are sent to the subagent for the line card. The subagent, in turn, sends the trap to the master agent on the control card, from which it sends a notification to the CLI, SNMP, or HTTP entity, depending upon the configuration (see Figure 8.6).

AgentX as defined by RFC 2257 is one example of an extensible agent paradigm for SNMP. In this architecture, a master agent uses SNMP to communicate with the outside world and relies on the AgentX protocol to communicate with other subagents. AgentX permits the subagents to register regions of the MIB for which they are responsible, so that the master agent can contact the appropriate agent for the MIB variable(s) or tables requested. Note that this permits the subagent to be completely isolated from the SNMP protocol since it does not have a direct communication with the external SNMP manager.

8.8 A Checklist for Multi-Board Software Development

When developing code that can be used in both single- and multi-processor environments, the design is more complex. State machines can be split, procedural interfaces need to be changed, pointer-based access is to be reexamined, and so on. The following provides a checklist for the development:

- Use messages as opposed to procedure calls between functional modules that can be split
- Split state machines and use messages for interaction between the state machines
- Use Message Abstractions and Name Services to handle both intra-board and inter-board communications
- Use Agent and Subagent methods for management

8.9 Evaluating the Single Control Card + Multiple Line Card Architecture

There are several advantages to the single control card + multiple line card architecture, including a single point of control for the control plane and management protocols, distribution of information like routing table information, relatively easy software migration from a single-processor architecture (if we use messages), and so on.

However, the architecture also has some disadvantages:

1. The control card is a single point of failure—if it goes down, so do all the control plane and management plane protocols.

2. The control card processor can become overloaded due to the large number of control plane tasks that need to be implemented. Some protocols such as OSPF and IS–IS require complex SPF (Shortest Path First) calculations, potentially tying up the control processor for long periods.

3. The control card–to–line card control interface is a potential bottleneck, since the line cards need to forward messages destined to the router on this interface to the control card.

One way to address these issues is to add additional control cards. Some systems follow such a scheme for scalability—for example, one control card could run all the interior routing protocols (OSPF, IS–IS and RIP) while another could run the exterior routing protocol like BGP and also the multicast routing protocols like PIM (Protocol Independent Multicast), MOSPF (Multicast Open Shortest Path First), and DVMRP (Distance Vector Multicast Routing Protocol). The line cards are aware of this division and send the received packets to the appropriate control card. Note that the protocols continue to have a complete system view since they communicate with all the line cards.

The power provided by the additional control cards comes with its price—the need for two control cards to exchange information adds complexity. The RTM needs to be common, since it will be used to determine the policy for the routes and route redistribution and pass this information to the IP Switching Task on the individual line cards. With the routing tasks split across multiple control cards, this becomes a challenge. Another issue to consider is the distribution of management function between the control cards.

Despite the complexity, this solution is more scalable. Hence, some designers use a completely distributed architecture for their multi-processor system, as discussed next.

8.10 Multiple Line Card, Fully Distributed Architecture

In this IPS implementation scenario, each line card has its own control processor (see Figure 8.7). The control processors on the line cards communicate with each other through a control interface on the backplane such as PCI or Fast Ethernet. There is no single control card. Instead, the control function is distributed among the control processors on the line cards. For simplicity, the management interface is present on one of the line cards, so there is a master agent on the line card and subagents on other line cards. Management PDUs are routed to the line card with the master agent.

Each of the control processors runs the complete set of routing protocols. They communicate over the control interface with each other, so they effectively look like multiple routers in a box, with the backplane control interface acting like a LAN. So the control cards actually run the routing protocols over the backplane LAN. Since the routing functionality is now distributed, the control processors are not as heavily loaded as a single control card.

Using Figure 8.7 as an example, assume that the SNMP agent is located on Line Card 3, and the external manager communicates with the router on Line Card 1. Also, assume the variable value requested by the SNMP manager is obtained from Line Card 2. The

SNMP packet is forwarded by Line Card 1 over the control backplane to Line Card 3. The SNMP agent on Line Card 3 communicates with the subagent on Line Card 2 to determine the value of the variable required by the external manager. This implies that the MIB views provided by the individual subagents are aggregated by the agent on Line Card 3, even though each of the Line Cards appears as a single router.

Figure 8.7   Fully distributed architecture.

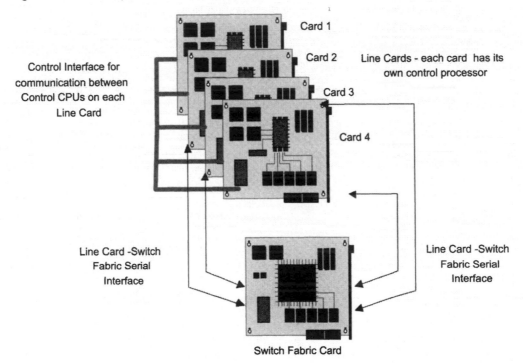

Card 1

Card 2

Card 3

Card 4

Line Cards - each card has its own control processor

Control Interface for communication between Control CPUs on each Line Card

Line Card -Switch Fabric Serial Interface

Line Card -Switch Fabric Serial Interface

Switch Fabric Card

While the distribution and MIB view abstraction is acceptable for tables, it is not as straightforward when we consider the routing protocols. If the OSPF protocol is to appear as a single entity to the external manager, the various instances on each of the line cards have to collaborate to provide the single OSPF instance view. This is conceptually similar to the split control plane view for OSPF "Hello"s, which involved two state machines cooperating through messages. The difference is that individual state machines are split across multiple boards but have to provide a single view for management.

Some systems avoid this problem altogether by providing the multiple OSPF instance view without having to combine the instances. However, the view is accessible only with the management agent and an instance ID—and not directly through the line card. This enables the administrator to configure the OSPF instances on a line card basis but at the same time not address them as individual routers. This is also called a *virtual router* approach. Some IP services boxes have used this architecture for scalability.

A variation on this is the *multi-router* approach, proposed recently. Here, an administrator can dynamically configure control and line card mappings, so that there are multiple physical routers in the same box. In this case, there is no need to have a unified management view for the routing protocols or cards, except to configure the mapping.

8.11 Failures and Fault Tolerance in Multi-Board Systems

The single control card + multiple line card architecture is the more common implementation for multi board systems in communications applications. It will be used as the basis for discussions on fault tolerance.

Multi-board systems are equally as susceptible to hardware failure as single-board systems. However, unlike the single-board systems, the multi-board architecture can handle failures by providing a switchover to another board. This is a requirement in carrier or service provider environments where multi-board systems are frequently deployed. The switchover is done to a *redundant* board, thus ensuring the *high availability* of the system. A redundant board is an additional board used to switch over to when the primary board fails. A system or network with high availability is able to recover from failures and continue operation.

This section covers common redundancy schemes used in multi-board systems for high availability and how software is modified to handle this.

8.11.1 Types of Failures

In a multi-board system, hardware failure can take multiple forms, the common ones being:

- Line Card Port Failure
- Line Card Failure
- Line Card to Control Card Link Failure
- Control Card Failure
- Switch Fabric Failure
- Line Card to Switch Fabric Link Failure

When a port on a line card fails, the line card identifies the failure and informs the control card, after which the protocol tasks on the control card process this as an event. For example, the RIP routing protocol needs to declare all routes reachable through that port as unreachable and propagate this information to neighboring routers. The FIB is recalculated by the RTM after the routing protocol has converged and then downloaded to all line cards. This FIB reloading at the line card could be incremental or complete, depending upon the impact of the port failure on the routes.

A line card failure or a card-to-card link failure manifests itself as a loss of heartbeat between the control and line cards. Specifically, the ICCP or messaging sub-layer on the control card detects that it has not received heartbeats from the line card in a configured time interval. The ICCP informs the control card CPU, which causes protocol tasks to be notified of the port failure on the line card. The resulting actions are similar to the single-port failure case.

The port failure and the line card failure cause only a degradation in system performance since the device still functions as a router, but with a lower port count. However, failure of the control card causes the system to lose the core of its intelligence—in terms of routing protocols, management agents, and so on. While the forwarding could possibly continue for some time using the information provided earlier by the RTM, the risk is that this information could be *stale*, causing misdirection of network data traffic and network instability. To address this, several systems provide control card redundancy to ensure continuous operation.

Switch fabric failure results in *islands of line cards*—they can forward packets between their ports but are unable to forward packets between the line cards. While the system is still functional, there is a severe performance degradation, so systems often have a redundant switch fabric card to address this. The line card–to–switch fabric link failure is similar to the switch fabric failure from the line card perspective, but the system is still able to function with just one line card being isolated.

8.11.2 Redundancy Scenarios with Control and Line Cards

There are two options for redundancy with control and line cards:

1. A redundant card for each card (1:1 redundancy)
2. One redundant card for N cards (1:N redundancy)

With 1:1 redundancy, each *primary* line card has an equivalent *backup* line card with the same configuration. When the primary card fails, the backup or redundant (or standby) card takes over. A highly available system requires that the switch from primary to redundant card take place without operator intervention. To accomplish this, the primary and backup card exchange heartbeats so that the redundant card can take over on both the switch fabric and control card link if the primary card fails. There are two options upon startup for the redundant line card:

Warm Standby.   The standby card was initialized in the redundant configuration and can request a download of the configuration from the system operator and continue operation. Warm-standby operations require operator intervention.

Hot Standby.   The configuration is obtained from the primary card, while it is still functional. The two cards do this by a periodic update from the primary to the standby card and/or when the configuration changes, also known as a checkpoint. When the standby card takes over, its information is as current as the last checkpoint from the primary.

The warm-standby operation is less flexible since the new configuration has to be provided to the redundant card—causing a disruption in system operation until the redundant card is fully operational. Moreover, this causes an extra burden on the system operator since the previous configuration has to be replicated step by step.

The hot-standby operation, on the other hand, requires three types of messages to be exchanged between the primary and redundant cards:

1. An initialization or bulk update, sent by the primary card when the redundant card comes up, provides a complete snapshot of the current configuration.

2. A periodic or on-demand checkpoint of configuration changes sent from the primary card to the redundant card.

3. A heartbeat message and response between the primary and secondary cards, sent in the absence of the checkpoint messages.

8.11.3 Control Card Redundancy

The card initialization, bulk update, and periodic checkpoints of a control card are shown in Figure 8.8. Note that when the redundant card comes up, it initializes itself, requests the complete configuration from the primary card, and then remains in standby mode. In this mode, it does not process events or messages other than the periodic checkpoint and heartbeat messages from the primary card. When it detects that the primary has not sent any heartbeats or checkpoints in a configured time period, it takes over as the primary card.

At this point, the software on the redundant card moves from a standby mode to the primary mode of operation. This causes the redundant card to start responding to all the standard events like queue events, timer events, and so on. The redundant card has taken over operation from the primary card.

This scenario implies that the software needs to operate in two modes: *primary* and *standby*. In primary mode, the software operates as before (non-redundant configuration), the only addition being that it will process messages from the standby card software and also provide initialization and checkpoint updates. In the standby mode, the software only obtains and updates the configuration using an initialization update and checkpoint updates.

Instead of each protocol task implementing the messaging scheme for the primary-to-standby transaction, system designers usually implement a piece of software called *redundancy middleware* (RM), which provides the facilities for checkpoints, heartbeat, and so on. The redundancy middleware offers a framework for the individual protocol and system tasks to provide information in a format appropriate for the standby. Protocol tasks make calls to the RM for passing information to the standby. The RM can accumulate the checkpoints from the individual tasks and send a periodic update or send the checkpoint immediately, which is usually configurable. The redundancy middleware uses the services of the ICCP messaging layer (see Figure 8.7) to communicate between primary and standby. On the standby, the RM provides the initialization or checkpoint data to the appropriate protocol tasks which have registered for this.

Figure 8.8   Control card redundancy.

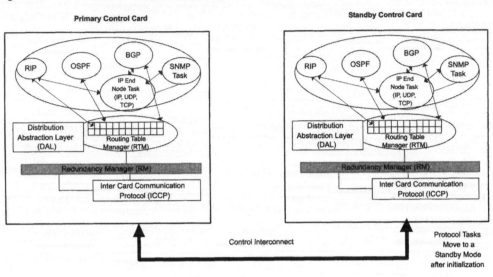

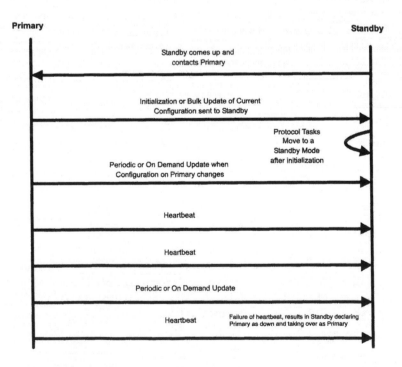

8.11.4 Line Card Redundancy

The 1:N redundancy scheme is used more often with line cards. It is less expensive in terms of hardware, since there is only one backup card for several line cards. Considering hot standby, the initialization and checkpointing activities are the same as the 1:1 case, except that the backup line card has to repeat initialization and checkpointing for every line card it backs up. This increases both the size of maintained data as well as software complexity since it now needs to track all the line cards. So, the preferred approach is warm standby, in which the card is initialized but the configuration is not loaded, since it will vary depending upon which primary fails.

8.11.5 Summary of Redundancy Model and Standby Models for Control and Line Cards

In the single control card + multiple line card architecture, only 1:1 redundancy is of interest. Each control card is backed up by a standby control card. Hot standby is more commonly used to avoid switchover delays. In line cards, 1:N redundancy with warm standby is more common.

Table 8.4 summarizes the various redundancy schemes for the control card and line cards.

Table 8.4   Redundancy schemes for control and line cards.

Card Type	Common Redundancy Model	Common Standby Model
Control Card	1: 1	Hot Standby
Line Card	1:N	Warm Standby

8.12 Summary

Multi-board architectures are required both for increased port densities and higher performance. It is preferred that the communications software is architected to run over both single-board and multi-board architectures. Using a message-based interface and facilities like a Distribution Abstraction Layer over an Inter-Card Communication Protocol can help in this regard.

The two common architectures for implementing a multi-board communications system are the Single Control Card + Multiple Line Card Architecture and the Multiple Line Card Fully Distributed Architecture. The former is more common while the latter is more complex but scalable. 1:1 and 1:N redundancy schemes with hot-standby and warm-standby configurations are common in multi-board architectures which need to provide high availability.

8.13 For Further Study

The ForCES Working Group in the IETF is looking at the details of the control and data element protocol. The Network Processing Forum (NPF) Software Working Group also provides some detail on the type of APIs on the control-forwarding interface. RFC 2257 provides more detail on the AgentX protocol. Cosine Communications (www.cosinecom.com) provides some information about the virtual router model. The multi-router model was first commercialized by Allegro Networks. Data Connections provides more information about a redundancy scheme used for MPLS. The Service Availability Forum (www.saforum.org) offers more information on redundancy and high availability models.

8.14 Exercises

1. Investigate some commercial RTOSes and list their support for multi board systems
2. Suggest some information that can be provided in a heartbeat message from the line card to the control card.
3. Another model often described in the literature is M:N redundancy. Explain how the 1:N and 1:1 models fit within the M:N redundancy model.
4. Provide some other examples where a split control plane model will be useful. What are the drawbacks of this model?
5. Investigate the ForCES model, and draw a diagram showing the presence of multiple control and multiple forwarding elements and how they are partitioned.

CHAPTER 9

Going About the Development

This chapter introduces some of the steps used by development teams for building communications software. It builds on the details of the previous chapters to outline the various steps during the design and development. The issues addressed include the stages of development, "build versus buy," and using a simulated environment as well as an OS Abstraction Layer. Tools of the trade like development environments and test equipment will also be covered.

9.1 Product Development Steps

This section outlines the typical steps followed in the design and development of software for communications devices. Individual document titles and formats may vary from company to company according to the software development methodology, but the concepts are the same.

Product development starts from two documents which are provided typically by the marketing and product management teams. These are the market requirements document (MRD) and the product requirements document (PRD). The market requirements document covers the target market, customer requirements, competitive analysis, revenue potential, product fit within company strategy, and so on. The product requirements document is derived from the market requirements document and details the features, feasibility studies, engineering tradeoffs, usability criteria, performance requirements, development and delivery schedules, and roadmap or projected life cycle. Some companies combine these into a single document.

The product requirements document forms a framework for the engineering team, which uses it as input to the high-level engineering design (HLD). The high-level engineering design details specific tasks and modules in the system at an architectural level, including functionality and interfaces. While the document is being prepared, an engineering team works on the system test plan, based on the features specified in the market and product requirements documents. The system test plan (STP) details the test strategy, test setup, tools and equipment required, test topologies, and so on.

The next engineering document is the low-level design (LLD) for each major module, which drills down deeper into the design, including data structures, state machines for the protocols, and so on. Once management approval has been granted to each of the above documents, the coding may begin. After coding and a code walkthrough process, unit testing is usually done on each module, followed by integration testing, which combines multiple modules to form the complete system. System testing is near the final stage of development, when the system test plan is executed.

The complexity of individual stages depends on the nature of the system. For example, unit testing for a simple buffer management scheme may be different from unit testing for a protocol. Similarly, unit testing of a hardware device driver may be more complex if the hardware is still under development.

The following sections cover the typical steps in the design and development of a Layer 3 IP Switch (IPS), using the software development life cycle given above.

9.1.1 Layer 3 Switch Product Requirements

In this phase, which follows the market requirements document, the product manager (in consultation with the engineering team) outlines the features that are required in the Layer 3 IP Switch (IPS) development, including the following:

- System requirements (single-board, multi-board, multi-processor)
- Operating system (Commercial UNIX, RTOS)
- Number and Type of Ports
- Software Protocols
- User Interface and Management (CLI, SNMP, CORBA, XML)
- Performance requirements
- EMI, Power requirements
- Standards Conformance (hardware, software, power, EMI)
- Phases of Development (Phased delivery schedule and expectations)
- Scalability (hardware and software designs to accommodate future expansion and scalability)

The product manager is the bridge between engineering, marketing, and the customer. The product manager plays the role of arbitrator in the event of a mismatch between external market demand and internal constraints. For example, the product may need to use the code from another product line for feature compatibility, or it might need an ASIC to be developed (to handle the performance specified). These constraints are clearly identified at the product requirements stage. Mature product organizations nail several of these issues up front, without the need to revisit them during development.

Build versus Buy

It is common in the product requirements stage for both development and marketing teams to discuss the "build versus buy" question. The engineering team can develop software modules such as protocol stacks from scratch or license the stacks from a third-party source code vendor or from a Linux distribution. Several equipment vendors prefer to license the stacks or components from vendors. This is more common with startup companies building new products. The decision of "build versus buy" is not restricted to software. Hardware designers have to choose between using merchant silicon like switching chipsets instead of developing their own ASIC.

Time to market (TTM) can be one of the most important considerations. Protocol stack software takes time to develop and requires a significant amount of testing for compliance and interoperability. The effort and time required for these steps may be significant. Moreover, since the protocols are standards based, there is very little scope for enhancement or differentiation, which can be another reason to license the stacks.

Development teams struggle with the issue of build versus buy since it requires a careful analysis of the tradeoffs. Table 9.1 provides a quick overview of the issues to be considered when choosing between internally developed code and third-party protocol stacks ("licensed code").

Table 9.1   Build versus buy issues.

Issue to be Considered	Code Developed Internally	Licensed Code
Time to Market	Longer—since effort & resources required to develop the individual protocols may cost more than the price charged by vendor for code	Shorter
Code Stability	Requires some level of testing until the code is stable and fulfills requirements	Tested and deployed widely—more stable
Complexity	Can be made less complex than third-party code if new code conforms to specific requirements of the system being built	Less or more complex depending upon the architecture—portability considerations may cause code to be more complex than normal
Performance	Optimization can be done for specific system for which it is being built	Cumbersome to optimize for specific system since it has been written to be portable to multiple systems
Cost	Engineering effort + testing effort—can be quite high	Dependent upon vendor pricing
Support	Developers support the code base—highly flexible but can be an issue with limited resources	Vendor provides support and updates to latest version of standards

Despite some of the advantages of licensing, engineers are divided in their opinion about the value of licensing code like protocol stacks from third-party vendors. The architecture of the licensed stack, its fitness for the system being designed, and the quality of the code are some of the factors that contribute to the decision.

9.1.2 High-Level or System Design

For the high-level design phase, the system is decomposed into individual tasks and modules, along with the interfaces between them. In the IPS, this consists of the following:

- Modular Decomposition into the individual routing tasks, IP Switching Task, Driver Details
- Internal interfaces between the modules—including interprocess communication for message passing between modules along with message types
- External system interfaces—including user interface and serial port connectivity
- Global Data Structures used in the system (common routing table) and access routines to manage these
- Events and notifications for each of the modules
- Provide a mapping to product requirements, such as routing table sizes, split control plane in multi-board systems

9.1.3 Low-Level Design

The low-level design (often called *detailed design*) is the second level of decomposition. Individual modules are broken into submodules with data structures, the interfaces between the submodules, and dependencies between the data structures. In the IPS, this consists of the following:

- Decomposition for individual tasks such as OSPF, IP Switching Task, drivers for the various types of interfaces
  - Depending upon the complexity, each of these modules may have its own detailed design document
  - If a third-party stack is licensed, the low-level design is a porting plan, in which the changes to the licensed software and its interfaces are specified.
- Key data structures in the individual tasks such as interface tables, neighbor tables, and protocol control blocks
- Pseudo code for key functions such as the main loop of the task, the shortest-path-first algorithm for OSPF/IS-IS
- Sizing and performance for each of the modules or tasks

9.1.4 Coding

This phase translates the detailed design into code. In the case of third-party stacks, it includes translating the porting design into code. Coding guidelines, as specified by the project, need to be followed for easy maintenance.

9.1.5 Testing

Unit testing can involve developing stubs for routines that are called by the individual modules being tested, test tasks in a multiple-task environment, and test code that drives the testing. Test packets and messages need to be constructed during this stage. The test packets could be constructed using packet generator test tools.

Integration testing can be done when individual modules are combined together in a phased manner. At each integration phase, the stubs are replaced by the individual routines in the interfacing module, while the test tasks are replaced by the newly integrated tasks. For example, during unit testing of the RIP task, engineers may stub out the interface routines to IP and have a test task that plays the role of the IP Switching task. A similar effort is required for the IP Switching Task. At integration, the "real" RIP task and IP Switching Task are tested together.

Test Plan

All types of testing stages require a test plan. This document specifies the following for each test:

- The scope of the test, including the objectives, and what is not included
- Test Tools to be used, including Analyzers, Packet Generators, external nodes
- Test Setup and data to be used
- Dependencies for the test
- Actual Test Procedure (connections, initialization, packet generation)
- Required Result
- Special instructions or comments

The format of this document will vary according to the organization, but it is important that all test cases include the information specified.

System Testing Criteria

System testing tests the product as a whole, after all the modules have been combined through integration testing. The specific criteria used for system testing are:

1. Functionality, including conformance
2. Stability
3. Scalability
4. Interoperability
5. Performance

Functionality and conformance testing indicates the testing to be performed to verify that the product satisfies the requirements specified in the product requirements document. With IPS, it can mean forwarding between all ports. Conformance indicates whether the product conforms to the individual protocol specifications and standards. RFC 1812, for example, specifies the behavior for an IP router. Conformance testing will verify that IPS functionality is in accordance with RFC 1812.

Stability testing is required to ensure that the product satisfies the reliability and quality requirements as specified in the product requirements. Generally this means that the application runs flawlessly for a specified period of time. The IPS can be tested for forwarding of packets between interfaces for extended periods of time to ensure that buffer and memory leaks do not occur.

Stress testing verifies that the system is able to perform correctly at "full load," both in terms of traffic and the configuration. In IPS, this can mean that the system has been configured with the maximum number of interfaces, has the maximum number of routes in its routing table, and is tested with traffic destined for multiple routes on all interfaces. In combination with stability testing, this is used to verify that the system performs correctly and in a stable manner.

Interoperability testing verifies that all components perform seamlessly and flawlessly together. This testing is key when standard protocols are involved. With IPS, it can include testing with third-party routers such as those from Cisco Systems or Juniper Networks. Since IPS needs to be deployed in a network which may include routers from other vendors, this is an essential part of the testing. This stage is less involved, if the protocol stack has been licensed from a vendor that previously verified interoperability with third-party routers. However, if the protocol was developed from scratch, interoperability testing can take longer.

Performance testing is used to verify the performance requirements specified in the product requirements document. For IPS, this can include forwarding rates between ports at full load, routing protocol convergence, time required to failover to a backup control card, and so forth. Performance numbers provided during this phase are used by marketing to showcase their product.

It should be noted that *testing is not linear but a highly iterative process*. Some test results such as scalability and stress testing may require a significant amount of time to fix the issues that caused the tests to fail. *Regression testing* requires that the entire set of tests be rerun after a fix—to ensure that the fix provided does not "break" any of the earlier functionality or tests.

9.2 Hardware-Independent Development

While developing communications software, engineers often need to develop and test the software even before the hardware is ready. It would be inefficient for the software team to wait for the hardware development to be completed. The schedule is drawn up with hardware and software development proceeding in parallel.

There are two ways that software can be developed and tested before implementing on the final hardware:

1. Using a simulated environment
2. Using a Commercial Off The Shelf (COTS) board similar to the final target

9.2.1   Using a Simulated Environment

The simulated environment is illustrated in Figure 9.1, describing the IPS running under a UNIX-based simulation environment. The IPS software runs on the simulator, with the driver handling the hardware ports, instead of running on the final target. The simulator provides the ability to create and delete tasks, pass messages and events between tasks and create timer facilities. The driver is a simulated module providing the same interfaces to the IP Switching Task as the final driver on the target. However, since there is no hardware involved, the driver will "receive" and "transmit" packets to another internal process called a "tester process."

The tester process is used to construct, verify, send, and receive packets to the simulated target. For example, a simple ping packet can be constructed and sent by the user process to the IPS. The IPS processes this packet as though it were received on a hardware interface and responds with a ping response. The driver sends the response to the user process, where it is received and verified. Scripts also provide a level of automation in the setup in which the packets are constructed, sent, received, and verified without user intervention.

In a more sophisticated test setup, control, configuration, and testing are completely external to the IPS simulator. For example, if IPS is a single UNIX process, a scheduler can be written within the process to schedule multiple threads (which are similar to tasks in the single–memory space environment of the process), so that the tasks execute on a simulated target. The driver module interfaces directly with the kernel instead of with an external user process. The driver module receives and transmits packets over the physical interfaces of the UNIX system, as shown in Figure 9.1. The advantage of this is that individual developers can work on their own UNIX workstations with a version of IPS running under the simulator. For example, a RIP developer can develop and test RIP using the simulator without dependency on the target hardware. Likewise, an OSPF developer can work on a simulator on his machine. This is also an advantage if there are a limited number of targets available for testing.

Commercial operating systems such as VxWorks™ offer simulators for development. The VxWorks™ simulator is called VxSim™, runs as a separate process under UNIX or Windows, and provides the same environment as the target. All system calls and services are supported under the simulator, so an application written to run under the simulator

Figure 9.1   A simulated environment.

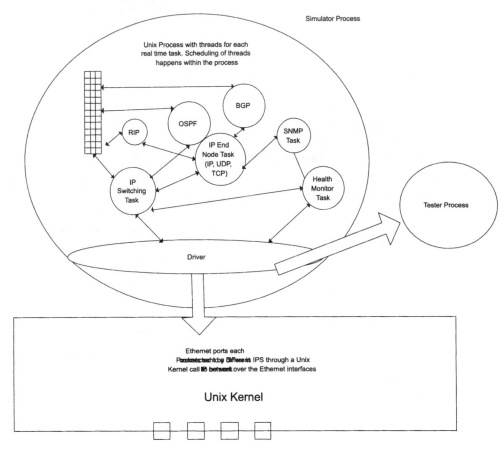

will not require modification when moving to the target. For example, calls such as `malloc` and `free` are supported under VxSim.

Not all simulators are the same. For example, with some RTOSes, memory protection between tasks may be available on the target RTOS but not on the simulator version. If IPS uses memory protection between tasks, then this feature cannot be tested on the simulator. Moreover, the system calls may need to be modified when moving from the simulator to the target.

9.2.2   Operating System–Independent Programming

Using an operating system simulator allows the applications to be largely invariant when moving from a host to a target. However, if the next-generation product uses a different operating system, the application will need to be modified to use the APIs and facilities provided by the new RTOS. This may not be a significant effort if the two operating systems have similar APIs. For example, there is less modification when migrating between two POSIX-compliant operating systems. However, where the interprocess communications mechanisms are significantly different, there can be an issue. For example, OSE 4.3 ™ uses message passing and a unique API for its IPC facilities, while VxWorks ™ uses POSIX-compliant APIs. When moving from VxWorks to OSE, each protocol that makes system calls will need to be modified, a significant effort in systems with a large number of protocol tasks.

A common way to address this is with an operating system abstraction layer, as shown in Figure 9.2. The OS Abstraction Layer (OSAL) provides a set of facilities that are available through API calls. These facilities are then mapped to the RTOS running the abstraction layer. For example, a call like OSAL_CreateTask will be translated to taskCreate in VxWorks and create_process in OSE. The OSAL has code for each RTOS over which it can run—as shown in Figure 9.2 where the code for OSAL_CreateTask has conditional code for each of the operating systems.

In an environment using OSAL, protocol tasks make OSAL calls instead of direct system calls. This ensures that they run without any change when moving to a different operating system as long as the OSAL has been ported to that operating system.

ACE OSAL

An example of an OSAL is the one available with the Adaptive Communications Environment (ACE) framework available from the Computer Science Department of Washington University, Saint Louis. ACE is an extensive framework incorporating IPC, memory management, timers, signals, thread management, Stream-based frameworks, and distributed communications services. The OSAL in ACE has been ported to multiple operating systems such as VxWorks™, LynxOS™, pSOS™ as well as desktop systems such as Windows NT™ and HP-UX.™ On a desktop operating system, the use of ACE can be an advantage for simulator testing. Instead of using a simulator like VxSim™, developers can run the protocol tasks under the ACE environment under UNIX. Functionality testing can be performed as before, and when ready to move to the target using VxWorks, developers can use the ACE framework for VxWorks and link it along with the protocol functionality.

Some companies have worked on the open source ACE and modified it to suit their own environment. This lets them optimize ACE for their own development like adding or removing features. Also, companies like Riverace offer technical support for ACE. So, developers can feel secure about the level of commercial backing and support when they obtain ACE from such companies.

There is an additional overhead involved with abstraction layers. This is due to translation of the abstraction layer call into a system call on the target OS. This necessitates that the OSAL be a functionally "thin layer." Another issue is the lack of OSAL standardization. Protocol stack vendors usually supply their own OSAL along with the protocol stack, while equipment vendors often build their own OSAL. Some effort is required to map the abstraction layers from multiple sources when building a system

Figure 9.2   Operating system abstraction layer.

Simulator Process

Unix Process with threads for each
real time task. Scheduling of threads
happens within the process

BGP

OSPF

RIP

IP End
Node Task
(IP, UDP,
TCP)

SNMP
Task

Tasks make call to
OSAL instead of
direct OS calls

IP
Switching
Task

Health
Monitor
Task

Tasks make call to
OSAL instead of
direct OS calls

Driver

```
OSAL_CreateTask (....)
{
#ifdef VXWORKS
    taskCreate (...)
#endif

#ifdef OSE
    create_process (...)
#endif

}
```

Operating System Abstraction Layer
(OSAL) - this is the only layer which
changes based on the OS. The
OSAL provides a wrapper for the OS
specific functions of:
—Task Management
—Memory Management
—Event Management
—IPC
—....

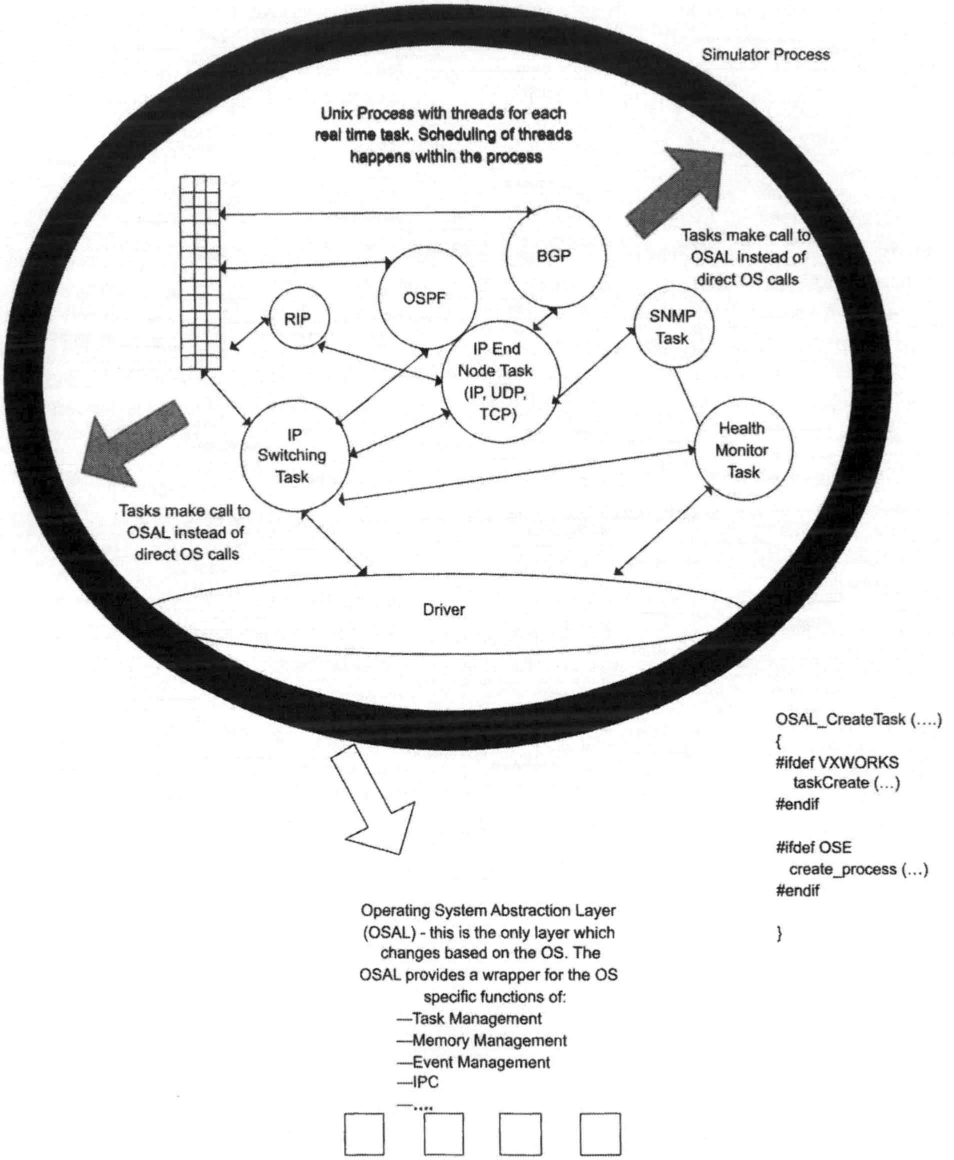

using third-party components. Despite these drawbacks, the use of OSALs is quite popular in communications software development.

9.3 **Using a COTS Board**

The acronym Commercial Off The Shelf (COTS) is often used in the embedded industry to indicate a product that is available commercially and can be used by multiple customers for their applications. Vendors such as Force Computer, Radisys, and Embedded Planet offer COTS boards for use by various vendors, including communications equipment manufacturers. Some equipment vendors integrate these boards into their own designs as-is, while others use these as an interim platform for development before the final hardware is ready. Our discussion will focus on the second scenario.

Assume that the final hardware platform for the IPS is a single-board, single-processor system using a Motorola PowerPC 750™, running VxWorks™. It will have 4 Ethernet ports and 2 Frame Relay (WAN) ports. The hardware designers will work on a cost-efficient design for the hardware, deciding on the components needed on the main board and via add-on cards. Details of the design vary based on the market and cost requirements. In the meantime, software developers can develop and test the IPS software on a COTS board using a PowerPC 750 processor running VxWorks. It is not likely that the COTS board will have the same number of interfaces required by the final hardware design, but it is a good platform to verify application functionality.

Using the COTS board lets engineers build the software using the same development environment and set of tools that will be used on the final target. This includes the compiler, linker, debugger, and the RTOS itself. Issues related to the processor and hardware can be sorted out during this phase. Since the COTS board often includes add-on PCI slots, developers can purchase COTS add-on cards (e.g., a PCI-based dual-port T1 interface for Frame Relay) and test the product extensively. Similarly, some security chip providers offer a validation platform for their chips using PCI cards that can run under VxWorks or Linux. These validation platforms can be added to the COTS board and used to verify the functionality of the silicon in the IPS environment as well as the performance of the IPS, using the security processor before its design into the final target.

Since the hardware interface support on the COTS board may be limited, developers need to maintain conditional compilation flags for those parts of the code baseline which depend on the hardware, such as a constant defining the number of hardware interfaces on the target. An include file specifying this variable would have the format shown in Listing 9.1.

Listing 9.1   COTS board include file.

```
#ifdef COTS_BOARD
    #define NUMBER_OF_HARDWARE_PORTS      4
#else
    #define NUMBER_OF_HARDWARE_PORTS 6
#endif
```

Teams need only a few COTS boards for development since these boards have limited use once the final target is ready. The boards can also serve as good validation platforms and, in some cases, as good demonstration platforms at trade shows or for early adopter customers wanting a proof-of-concept platform.

9.4 Development Environments and Tools

Depending upon the processor, developers will need to license compilers or cross compilers that run on their host workstations. For example, a developer working with a VxWorks™ development environment under Windows™ for a target using the PowerPC™ 750 processor will need a cross compiler, linker, and loader running under VxWorks for the PowerPC processor. RTOS vendors may offer these tools as part of the development environment, such as the Tornado Development Environment from Wind River. Alternately, the developer may need to license the tools from a third-party vendor such as Green Hills.

Some RTOS vendors offer the source code for their operating systems as part of the development licensing agreement. This is more the exception than the rule. Most vendors offer only the object code for the operating system for the processor needed by the customer. In our example, Wind River will provide VxWorks object files for the PowerPC processor, which need to be linked into the application comprising the protocol tasks, OSAL, and management tasks, as shown in Figure 9.3(a). Some system files include a few source files and include files which can be modified for the specific target hardware initialization, application startup, and timer interrupt handling.

The cross-development tools for the PowerPC 750 are used to compile the source code for the application into PowerPC object code and linked with the VxWorks OS object files to build a loadable image to download to the target. The download happens through an Ethernet port on the target, which is designated as the management port. The image can also be burned into flash for a faster boot. For multi-board systems, a separate image is burned into flash on each board.

On startup, initialization status is displayed with diagnostic messages on a serial port. The messages can be viewed by connecting a terminal or a PC with a terminal emulation program with parameters set to match with those for the serial port on the board. The messages include initialization status for DRAM, SRAM, hardware controllers, and the creation of data structures and buffer pools. Once initialization is complete, each of the tasks wait on events in their main loop.

To debug on the target, developers need to compile with the debug option. The code can be debugged through the serial port or through a host-based debugger which connects with the target over an Ethernet or serial connection. A target server exists on the target to act as the "local representative" of the host-based debugger (see Figure 9.3 (b)). The sequence of operations is as follows:

1. The developer uses host-based GUI commands for debugging.
2. The debugger communicates to the target server.
3. The target server performs the debug operations as though it were locally attached.
4. The target server sends the results of the debug operation to the host-based debugger.

Figure 9.3   Typical developnent environment.

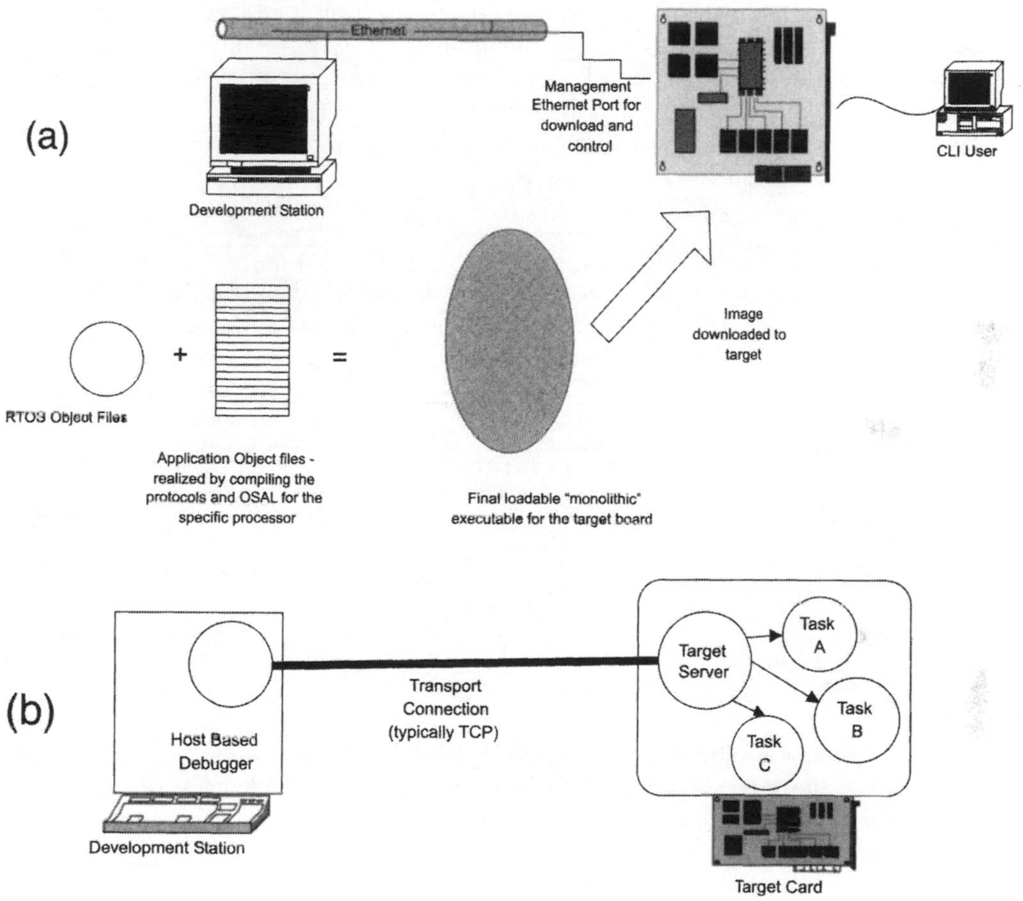

(a)

Development Station

Management
Ethernet Port for
download and
control

CLI User

RTOS Object Files

+

Application Object files -
realized by compiling the
protocols and OSAL for the
specific processor

=

Final loadable "monolithic"
executable for the target board

Image
downloaded to
target

(b)

Host Based
Debugger

Development Station

Transport
Connection
(typically TCP)

Target
Server

Task
A

Task
B

Task
C

Target Card

Several RTOSes offer file system capabilities either in local RAM or a remote development station. In this case, it may be possible to mount a directory on a user development station as a file system on the target; then diagnostic information can be written to a file on the target system. The information is transmitted using a UDP or TCP connection from the target to the host.

9.5 **Test Tools and Equipment**

Testing the communications software on the target involves various types of test tools and equipment. For example, protocol conformance testing for the IP suite can be done with a tool like Automated Network Validation Library (ANVL) from Ixia Corporation.

The test generates packets for specific protocols on an interface connected to the router, while the other interface is used to observe behavior (see Figure 9.4). For example, to verify correct forwarding, the tool can generate a packet on Port 1, with a destination address equal to the IP address of Port 2 in Figure 9.4. The system under test (SUT) will forward the packet on its Port B to Port 2 of the test tool. The tool receives the packet, verifies it, and updates the test result information.

Figure 9.4   ANVL test tool.

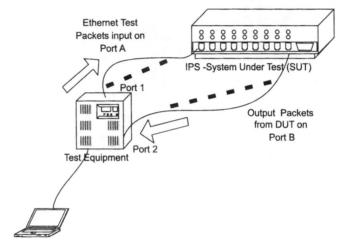

Test Result Monitoring via GUI interface to Tester

While ANVL is a conformance tool offering some level of performance-based testing, there are specific types of test equipment such as those from Agilent, NetTest, and Spirent which incorporate conformance as part of a test suite portfolio. These types of test equipment perform two basic functions:

Analysis   the test equipment captures the frames and then decodes and analyzes the data for correctness.

Simulation   the test equipment generates packets for specific protocols and verifies the responses.

Simulation often involves a light-weight implementation of the protocol with its state machines, packet generation and reception, and a limited amount of configuration.

The test equipment can also be used to validate performance such as the forwarding rate, latency, jitter, and so on. This is usually done on the complete system with the hardware and software subsystems integrated.

The Importance of Testing

Testing is a very important part of the product development life cycle. Communications equipment needs to be tested with other types of equipment in complex network topologies, since customers may deploy the equipment in large networks. Several equipment vendors build large labs for testing their equipment, in some cases doing a mini-replica of customer topologies. In addition, there are several test laboratories like the University of New Hampshire, George Mason University, EANTC, who offer testing services. Equipment and software vendors can take their products to these facilities to get them tested and certified.

Due to the complexity and interaction of multiple protocols as well as the high data rates, it is important that testing be carried out in a structured manner at all stages— unit, integration, and system testing. Like other areas of embedded systems, inadequate testing of communications software can come back to haunt the developers. A problem in the field is much more difficult to fix than one in the lab.

9.6 Summary

Product development follows a sequence of steps starting with the Market Requirements Document to the Product Requirements, High-Level or System Design, Low-Level Design, Coding, Unit, Integration, and Testing. The steps may vary according to the organization, but there is a significant amount of process involved here.

Developers need not wait for the hardware to be ready to test their software. Simulators, either home grown or provided by the RTOS vendor, are often used. An OS Adaptation Layer is often used to isolate the communications software from the underlying RTOS so that it can be easily ported to multiple operating systems. COTS boards, along with development tools, can be used for the initial hardware-based testing. Finally, testing can be performed using several commercially available test equipment/tools via analysis and simulation.

9.7 For Further Study

Washington University at St. Louis provides more information about ACE. COTS board vendors like Force Computer and Embedded Planet furnish details about their offerings on their Web sites. Test equipment vendors like Ixia describe their various products via seminars and on their Web sites.

9.8 Exercises

1. Enumerate and list some of the considerations in choosing an RTOS for a communications system.
2. List some limitations of simulator-based and COTS-based testing.
3. Explore the types of test scripting support provided by test tool vendors.

Appendix A

Examples from Commercial Systems

This appendix will use commercial systems to help illustrate some of the hardware and software architectural points discussed in the earlier chapters. This is intended only to outline the key features based on public information (feature sheets available on the manufacturer Web sites and books) and not provide a detailed description.

In keeping with the common thread in this book, we will look at IP Routers and consider two manufacturers: Cisco Systems and Juniper Networks. The two products considered are core routers, i.e., multi-board systems. Readers are encouraged to look up the Web sites of these vendors (www.cisco.com and www.juniper.net) for up-to-date information about their products. In addition, readers should also check out the information provided by other router manufacturers like Avici Systems, Extreme Networks, Foundry Networks, Netgear, Nortel Networks, and Riverstone Networks.

Juniper Networks M-Series Routers

Juniper Networks first started out making routers for the core of the Internet, where high data rates and large traffic volumes are the most important considerations. The company has subsequently branched out into edge routers, further strengthened by its acquisition of Unisphere Networks in 2002. The specific focus in our discussion will be on the M-series routers running the Junos™ software.

The M-series routers range from the M5, with 5 Gbps throughput, to the M160, with 160 Gbps throughput. For illustration, we will consider the M160, which has 2 chassis per rack and 32 PICs (Physical Interface Cards) per chassis and also provides switch fabric and routing engine redundancy. This type of router is an example of a multi-board system using the Single Control Card + Multiple Line Card Architecture. The Routing Engine is the control card running the Junos™ software, and it uses an Intel-based PCI platform. It uses a dedicated connection to the Packet Forwarding Engine (similar to the switch fabric) for changing the forwarding tables used by the PFE for switching traffic.

The Junos™ software is based on a protected memory design in which one process cannot corrupt another. This facility is also useful for in-service upgrades of specific software modules, a feature provided by the fact that Junos is not a "monolithic code base."

The key processes in the system include:

1. Routing Protocol Processes
2. Interface Process—to configure and control the various interfaces
3. SNMP and MIB-II Process—for management from an external SNMP manager
4. Management Process—for managing the other processes in the system, including the CLI and restarting processes that have failed
5. Routing Kernel Process—used for communication with the Packet Forwarding Engine regarding the forwarding table changes

The architecture is similar to IPS except for the presence of additional processes like the Interface Process and the Management Process. A multi-board implementation of IPS would need to have processes similar to the ones in Junos.

For management, Junos offers both a CLI- and an XML-based API called JunosScript.

Apart from these, the router can also be configured via SNMP.

Cisco Systems 12000-Series Routers

Cisco Systems has the largest range of routers in the market today. The Cisco 12000 product line is the one of interest for this discussion, since it is the closest competitor to the Juniper Networks M-series routers. The 12000 series starts from the 12008 router, with a throughput of 40 Gbps, to the 12416 router, with a throughput of 320 Gbps. Considering the 12410 router, which has a throughput of 200 Gbps, we note that it has redundant route processors and switch fabrics, as well as redundant power supplies.

The architecture is also based on the Single Control Card + Multiple Line Card architecture. The line card–to–switch fabric communication is in the form of fixed-size cells. The control card–to–line card interface is via a 1 Mbps serial bus called the Maintenance Bus.

The control card (called the Gigabit Route Processor) runs the Cisco IOS™ (Internetwork Operating System) software. The CPU used is a MIPS family processor. The switch fabric is also known as the forwarding engine This software is used across multiple Cisco routing products but has different profiles based on the product line.

Cisco IOS™ Software

IOS is both the operating system and the software required for the various protocols. In that sense, it is a monolithic piece of code without a separation between the application and the RTOS. IOS provides the capability for both CLI and SNMP management. The CLI can be accessed via a serial port and via telnet.

IOS uses the term "process" to indicate threads and associated data (similar to tasks specified in our description). Processes perform various functions like switching packets, system maintenance, and implementing routing protocols. For use in partitioning data according to memory (SRAM, DRAM, and so on), IOS uses the concept of *regions*. Free memory is managed via a series of memory pools

Packet buffers are managed via a buffer pool manager. Packet buffer pools are allocated out of memory from memory pools.

Unlike most RTOSes, IOS uses a run-to-completion model of execution. Processes cannot preempt each other, so a process has to relinquish control for another process to run.

IOS device drivers provide a hardware abstraction layer between the hardware and the OS. The interface is via a data structure called the *Interface Descriptor Block* (IDB), one for each interface.

For more details, the reader is referred to "Inside IOS Software Architecture" by Vijay Bollapragada, et al., Cisco Press, 2000.

Glossary of Common Terms and Acronyms

1:1 redundancy A high-availability scheme in which each device is backed up by a redundant device. In a system with 1:1 redundancy for line cards, each line card would have its own backup line card.

1:N redundancy A high-availability scheme which N primary devices are backed up by a redundant device. In a system with 1:N redundancy for line cards, N line cards would have one backup line card.

Ack Another name for Acknowledgement. Sent by a receiver of a packet to the sender.

action routines The part of a state machine which implements the action for a specific event in a specific state. Action routines may be shared, so you could have the same action routine invoked for several state–event combinations.

agent A software entity on the embedded device which processes and responds to manager requests. An SNMP agent is one example.

AgentX Agent Extensibility as defined by RFC 2741 from the IETF. The architecture defines a master agent, which processes SNMP messages, and one or more subagents, which interact with the master agent for management operations. The subagents do not need to deal with SNMP; they use only the AgentX protocol to interact with the master agent.

ANSI American National Standards Institute. A private body based in the US which specifies its own standards. The body also participates in ISO standardization activities.

API Application Programming Interface. The specification of the services provided by a module for use by other modules, including the method of invoking the services.

ARP Address Resolution Protocol. A protocol used in the IP world to determine the IP address–to–MAC address mapping.

ASIC Application Specific Integrated Circuit. A custom IC for a specific function. ASICs are usually designed internally for a specific product. They are expensive to design but are optimized for the required function.

ATM Asynchronous Transfer Mode. A cell-based, connection-oriented technology with 53-byte cells.

authentication The method of verifying the sender of a message.

backplane A hardware interconnect in multiboard systems that which individual cards plug into. Each card or blade communicates with other cards over this interconnect.

basic parameters The minimum set of parameters required to configure a protocol/device. These parameters cannot take default values and need to be explicitly configured.

BGP Border Gateway Protocol. An exterior gateway protocol used for exchanging routes in the Internet.

blocking call A function call which does not return until the complete operation is performed in the implementation. The call waits or "blocks" until the operation is performed. An example is a call to receive a packet, which does not return until the packet is actually received.

board support package (BSP) A port of a Real Time Operating System to a specific hardware board. The package will include processor-specific code, including initialization, checks for the DRAM and other devices on the board, and hooking up to timers. The BSP for several COTS boards are provided by the RTOS vendor.

bps Bits per second.

bridge A Layer 2 device which learns the location of nodes in its connected LANs by using the source address from received frames. A variation of the bridge is the Layer 2 switch, which can provide bridging between multiple independent ports at the same time, often using hardware support.

broadcast address A destination address value which indicates that the packet is to be delivered to all addresses in the network.

buffer A temporary memory block used to store data during processing. It is used to transport payload and control information within a communications device.

buffer management The techniques used to handle the allocation, freeing, manipulation, and control of buffers in the system.

caching A method that uses a temporary memory block or area to store recently used information.

CAM Content Addressable Memory. A memory mechanism in which a search is done in parallel across all locations in the memory.

chained buffer A buffer which exists as part of a linked list of buffers. The entire buffer chain is used to house a single packet.

checkpointing A technique used to indicate the current state and configuration of a primary system to its peer redundant system. If the primary system fails, the redundant system can proceed with the information from the last checkpoint.

checksum An integer value used to verify that the packet is uncorrupted.

chipset Two or more integrated circuits designed to work together.

circuit switch A switching technique in which a dedicated circuit is established between source and destination.

CLI Command Line Interface. A text-based user interface for configuring and controlling a device. Also known as Man–Machine Interface (MMI).

conformance testing Testing done to indicate that the product conforms to standards. Protocol conformance is one type of conformance. A protocol like PPP can be tested for conformance to the various PPP RFCs.

connection oriented A method which requires the setting up of a connection between two nodes before traffic can be sent between them. TCP is an example of a connection-oriented protocol.

connectionless A method which does not require the setting up of a connection between nodes for data traffic. Each packet contains the address of the destination node. IP and UDP are examples of connectionless protocols.

context switch An operating system action which causes the suspension of the currently executing task and begins the execution of another task. The context switch takes a finite amount of time due to the saving of one context (program counter, stack and contents of registers) and loading of another.

control block An anchor or root data structure used to access all the other information related to a protocol, interface, or infrastructure software module, including configuration, control, status, and statistics. A control block can be used in other contexts, say, for a protocol (PCB), hard-

ware interface (HICB), or protocol interface (PICB).

control card A hardware board in a multi-board system which runs the control protocols to communicate with peers and construct tables for use by the line cards.

control plane In the classical planar networking architecture model, this is the function which communicates with peer entities to build up tables for use by the data plane. Protocols like UNI 4.0 ATM Signaling and OSPF are in the control plane.

controller A hardware device to perform the specific functions required for a network interface, e.g., an Ethernet controller.

CORBA Common Object Request Broker Architecture. An architecture specified by the Object Management Group (OMG) for distributed object representation and utilization.

COTS Common Off The Shelf. Typically used to describe hardware or boards which are available from vendors without any need for custom development.

CRC Cyclic Redundancy Check. Inserted as a verifier by the sender into a frame. The value is calculated on the data in the frame. The receiver verifies the integrity of the transmission by calculating its own CRC on the received frame and comparing it with the CRC in the frame.

DAL Distribution Abstraction Layer. A component used to hide the distribution of functions across multiple boards.

data plane In the classical planar networking architecture model, this is the core function of a system. Functions on this plane provide data forwarding, switching, and packet processing based on the information built by the control plane.

demultiplexer The software that uses the information in the packet header to direct the packet upward through one or more protocol stacks.

detailed design The stage of a software development project in which the high-level modules are broken down into submodules, internal and global data structures are specified, and the processing logic outlined, where applicable. This is the phase before the coding phase.

device driver The software used to permit the applications to utilize a device without knowing all the details of the hardware.

differential timer block A timer block which stores only the difference in timer ticks from its previous block in the timer list. If there are two timer blocks required with one to time out in 10 seconds and the other in 15 seconds, the first block will have a timer count of 10, and the second will have a timer count of 5.

distance vector protocol A type of routing protocol which advertises reachability information to its immediate neighboring routers, based on which they construct their routing tables. Examples are RIP and IPX RIP. Contrast with a link state protocol like OSPF.

DMA Direct Memory Access. A technique used to transfer data directly between the network controller and memory without processor intervention.

DRAM A generic term used to describe various types of Dynamic Random Access Memory, including Synchronous DRAM (SDRAM), Double Data DRAM (DDR DRAM), Rambus DRAM (RDRAM) and Quad Data Rate DRAM (QDR DRAM).

DSL Digital Subscriber Loop. Used for high-speed communication in the "last mile," which is the link from the service provider to the subscriber (enterprise or home). Variants include ADSL (Asynchronous DSL), SDSL (Synchronous DSL), and their enhancements.

DVMRP Distance Vector Multicast Routing Protocol. A protocol specified by the IETF for multicast routing in the Internet.

dynamic allocation An allocation scheme in which the assignment is done dynamically.

edge routers Routers used on the "edge" of the service provider network. These routers interface via WAN links to the various customer premise routers. Due to the large number of customer connections, these boxes are usually high end/chassis based, with a large number of interfaces.

EEPROM Electrically Erasable and Programmable Read-Only Memory.

encapsulation A technique in which successive layers add a header to the outgoing packet.

encryption The process of transforming data such that only the intended receiver of the data will be able to read it.

end node A node which originates and terminates traffic in a communications network.

Ethernet A Local Area Network (LAN) technology using CSMA/CD. Ethernet can operate at 10, 100, 1000, or 10000 Mbps.

Ethernet address A 6-byte value uniquely identifying an Ethernet interface. Ethernet frames use the destination and source

Ethernet addresses at the beginning of the frame.

event A form of notification which is to be acted upon. There are several categories of events: timer events, message-queuing events, internal or local events, frame reception events, and so on.

fabric Also called a switching fabric; a hardware mechanism used to interconnect multiple functional units such as the interconnection among the input and the output ports.

fast Ethernet Ethernet technology that operates at 100 Mbps.

fast path The path that a normal/frequently encountered packet follows in the system implementation (hardware or software).

FIB see Forwarding Information Base.

flash A read–write semiconductor device used for storage of code and/or data. Code can be run out of flash.

flooding A technique where an incoming packet is copied and sent to all outgoing interfaces.

flow A group of packets of one general type.

flow control A set of techniques in which a receiver is able to moderate the rate of a sender.

ForCES Forwarding and Control Element Separation. A working group in the IETF which attempts to standardize the framework and associated mechanisms for the exchange of information between a control (plane) element (CE) and a forwarding (plane) element (FE).

forwarding information base (FIB) The subset of routes extracted from the Routing Information Base (RIB) by the RTM and distributed to the various forwarding elements.

fragmentation The process of dividing a packet when its datagram exceeds the MTU (Maximum Transmission Unit) of the network over which it is traveling.

fragmentation/reassembly Used in IP where messages larger than the MTU (Maximum Transmission Unit—i.e., max packet size) of an attached network are divided into smaller packets called fragments and sent towards the destination. The destination IP layer reassembles the various fragments and provides it as one single IP packet to its higher layer.

frame The name given to the packet recognized by the hardware.

frame relay A connection-oriented, public packet–switched service offered by the service provider.

FTP File Transfer Protocol. A protocol used for file transfer between hosts in the TCP/IP world. FTP runs over TCP.

functional interface Same as Procedural Interface. An interface between two modules or submodules where function or procedure calls are used, as opposed to sending and receiving messages between them.

gateway A internetworking device capable of relaying user information among networks with different architectures and protocol suites. Another name for a router.

gigabit Ethernet Ethernet technology that operates at 1 Gbps.

GPP General Purpose Processor. Used to describe a processor which is able to run general-purpose code. Examples include common control plane processors like the

MIPS and PowerPC line. In contrast, a network processor is a special-purpose processor.

GVRP Generic VLAN Registration Protocol as defined in IEEE 802.1Q for updating VLAN membership information between switches. Switches send GVRP messages between each other to update the membership information.

hashing The process whereby an algorithm is used to distribute evenly into a number of possible buckets in a hash table.`

HICB Hardware Interface Control Block. A root or anchor block for accessing the configuration, status, and statistics of a hardware interface.

high availability A feature wherein the system is available for most of the time. A system usually implements high availability using various schemes. Using redundant components is one scheme.

high-level design Often known as System Design. In this phase, the software system is broken into modules and tasks. This phase also specifies the interfaces between the various modules and tasks, along with global data structures.

host An end node in a network. Hosts communicate with other hosts, sending their traffic through switches and routers if the other entity is on a different network.

hot standby A redundancy scheme in which the standby card is initialized and receives updates from the primary card even while the primary is functional. This permits the standby card to take over very quickly when the primary card fails.

HTTP HyperText Transfer Protocol. An application layer protocol for accessing web pages. HTTP runs over TCP and is often used to manage a communications device via a browser. See Web-based management.

ICCP Inter-Card Communication Protocol. A protocol used for communications between the control card and line cards on top of the interconnect.

ICMP Internet Control Message Protocol. Used for diagnostics and control messaging in the IP world.

IEEE Institution of Electrical and Electronic Engineers. A body which also sets standards in various areas of networking. The IEEE 802 committee is the most relevant for networking. IEEE 802.3 is the Ethernet standard, while IEEE 802.11b, 11a, and 11g are the Wireless LAN standards.

IEEE 802.1D The IEEE standard specifying the Spanning Tree Protocol for Layer 2 bridges.

IEEE 802.1Q The IEEE standard specifying the VLAN registration protocol (GVRP) for exchange of VLAN registration information between bridges.

IEEE 802.3 The IEEE standard which specifies CSMA/CD-based Ethernet.

IETF Internet Engineering Task Force. The body responsible for the specification of the protocols used in the Internet.

inter-process communication (IPC) The mechanisms used for process-to-process communication. These can be determined by the operating system and can include message queues, mail boxes, shared memory, and so on.

interface A term used to describe the layer of interaction between two modules and the procedures used on this layer.

interface Used to describe a physical port of a system. See also Physical Interface and Logical Interface.

interop testing Testing for determining the interoperability of the equipment and software with other vendor products. Generally, a requirement before a networking product is shipped.

interrupt processing The opposite of polling. Processing occurs only if there is an interrupt, so there is no wasting of cycles due to polling.

interrupts Events that tell the operating system to stop its current activity and take action.

IP Internet Protocol. The protocol used in the Internet. It was originally specified by the IETF in RFC 791.

IPS Acronym for an IP Switch. In the text, this is the Layer 3 IP Switch/Router—a device which performs IP forwarding.

IPSec IP Security. A protocol used in IPv4 and IPv6 for securing IP communications between a sender and a receiver. IPSec involves both authentication and encryption.

IPX Internetwork Packet Exchange. A Layer 3 protocol specified by Novell for communication between its Netware servers and clients. Several IPX networks have been deployed in enterprises.

ISDN Integrated Services Digital Network. A set of protocols used to send voice and data over a network in digital format. Calls need to be set up before the voice or data transfer can take place.

IS–IS Intermediate System–to–Intermediate System routing protocol developed originally by the ISO. It has been enhanced by

the IETF for IP networks. The protocol is conceptually similar to the OSPF protocol in that it uses the Djikstra shortest-path-first (SPF) algorithm for calculating routes.

ISO International Standards Organization. A body which publishes a set of standards for networking.

ISR Interrupt Service Routine. The code that runs when the interrupt occurs.

ITU-T A subcommittee of the International Telecommunications Union, a global body that drafts technical standards for all areas of communication. ITU-T has specified standards for SS7, X.25, and ATM.

kernel The core of the operating system which implements system calls and provides access to memory, disk, and devices.

kernel mode Used to indicate execution of a function in a supervisory mode, where it has access to all kernel functions and devices.

LAN Local Area Network. The most popular one is the Ethernet LAN specified via IEEE 802.3.

Layer 2 Used to indicate the Data Link Layer, and, in the case of LANs, a MAC address based function, which deals only with flat networks without routing.

Layer 2 switch A MAC bridge over which the forwarding between ports A and B can proceed independent of the forwarding between ports C and D. Forwarding is typically done in hardware to enable this.

Layer 3 switch An IP router which typically uses hardware-based forwarding.

line card A component of a multi-board system which provides its physical ports. The ports are housed on media modules which plug in to line cards or on the line cards themselves.

link   A physical connection between two nodes in a network.

link handler   A component of the OSE real-time operating system used for distributed messaging.

link state protocol   A type of routing protocol which floods status updates of connected links throughout the network for each router to build a database of the complete topology. The routers can then apply the SPF algorithm to determine the shortest path to various destinations. Examples are OSPF and IS–IS. Contrast with a distance vector protocol like RIP.

logical interface   A term used for the abstraction of an interface. A logical interface can have a one-to-one mapping with a physical port (e.g., a serial port) or correspond to one of several interfaces on a physical port, e.g., a set of PVCs on a Frame Relay serial port.

longest prefix match (LPM)   A table lookup scheme used in IP routing in which the prefix match covers the most bits.

low-level design   Same as detailed design.

low-water mark   An indicator that the number of entries in a queue/memory or buffer pool has fallen below a predetermined level.

MAC   Medium Access Control. The data link layer sublayer which is responsible for access to a shared medium. Typically used to specify the access mechanism of LANs like Ethernet.

MAC   Media Access Control. A reference to the Layer 2 protocols used to access a network.

MAC address   Used as a synonym for Ethernet address.

management plane   The third part of the classical planar networking architectural model.

manager   An entity responsible for the control and configuration of various devices in the network. It can be a remote entity like an SNMP or CMIP manager or a local one like a CLI user. The manager sends requests and obtains responses from the communications device.

market requirements document (MRD)   A document generated by Marketing to provide specifications/market requirements for a product. This is the first step in the product development cycle.

master agent   Defined in the AgentX architecture, a master agent provides the management functionality for a system. It is typically on a control card and interfaces with the subagents on the line cards for management operations. The master agent is the only one which talks directly to the manager. The subagents communicate only with the master agent—the protocol used on this interface is AgentX.

memory management   A set of facilities usually provided by the operating system which allows applications to allocate, release and manipulate memory blocks for their functioning.

memory protection   Usually implemented on processors which have a memory management unit (MMU), this ensures that two areas of memory are protected via the MMU. So, if code in one area attempts to access the other, a memory violation exception will be raised by the MMU.

messaging interface A scheme in which entities desiring to communicate with each other use messages. The messages are filled in by the source entity and passed on to the destination entity.

MIB Management Information Base. Defines the network-related variables that can be read or written on a node.

MIB-II The Management Information Base specified in RFC 1213. This includes the configuration/control of interfaces, IP and TCP protocols, SNMP, and forwarding and address resolution tables. This is the standard which most IP routers and end nodes implement.

monolithic control plane An implementation in which the control plane functions like routing, and signaling protocols are fully implemented on the control card. There are no control plane functions on the line card, unlike a Split Control Plane implementation.

MOSPF Multicast OSPF. An extension of OSPF routing protocol to handle multicast routing.

Motorola PowerPC A popular RISC processor from Motorola used in several communications equipment designs.

MPLS Multi-Protocol Label Switching. A connection oriented technology used to forward traffic based on labels.

MTU Maximum Transmission Unit. A value indicating the maximum size of a packet/frame that can be sent on a specific interface.

multi-service switch A switch which forwards various types of traffic between interfaces. For example, the same device can implement switching of ATM cells, Frame Relay packets, and Ethernet frames. It may also switch TDM frames.

multicast A method of transmitting packets to a subset of nodes on a network.

multiplexer A device that combines distinct channels or streams into a single stream.

mutual exclusion An agreed-upon way of accessing a shared resource when multiple entities are involved. The entities agree that only one of them can access the resource at a given time.

network address translation (NAT) A mechanism used for translating between local IP addresses and global IP addresses for extending the IP address space.

network byte order Similar to the big-endian method of representation, in which the transmission of integers across the network begins with the most significant byte.

network prefix Each IP address is composed of a network prefix and a host prefix. The network prefix identifies the network that the node with this IP address belongs to.

network processor A programmable processor which is optimized to process packets at a very rapid rate. The instruction set handles only network functions. Programmers can write code using this optimized instruction set and download to the processor for handling packets.

NIC Network Interface Card. Another name for an add-on adapter card for networking purposes, e.g., an Ethernet NIC for a PC.

node A term used to describe a communicating entity on a network. A host is an end node, while a router could be a network node or routing node.

non-volatile RAM (NVRAM) Small IC powered by a battery, used to store system parameters.

NP Network Processor.

NPU Network Processor Unit. Another name for a network processor.

Nucleus A real time operating system from Mentor Graphics.

OSAL Operating System Adaptation Layer. Used to isolate the application from the underlying operating system. An application makes only OSAL calls, so it can be ported to multiple OSes as long as there is an OSAL for that OS.

OSE™ A real-time operating system from OSE Systems. It relies primarily on message passing for its IPC.

OSI Open Systems Interconnect. The seven-layer model for networking specified by the ISO.

OSPF Open Shortest Path First. A routing protocol based on a link-state algorithm, in which every node constructs a topography and uses it for forwarding.

packet The unit of transmission in a packet-switching network. The complete message is divided and sent as individual packets through a network.

packet switch A device which forwards or switches packets through a network.

payload The data carried in a packet.

PBX Private Branch Exchange. A circuit-switching device which is used to switch calls within the same premises.

PCI Peripheral Component Interconnect. A bus to interconnect processors and peripherals.

PDU Protocol Data Unit. A term used to describe a packet.

PDU preprocessing A set of operations in which a PDU is validated, verified and classified so that the appropriate event can be sent to a protocol state machine.

peer The corresponding node for the other end of the communication. The peers communicate with each other using a protocol.

PHY A term used to describe the physical layer device or chip interfacing to the line.

physical interface A physical port in the communications system.

PICB Protocol Interface Control Block.

PICMG PCI Industrial Computer Manufacturers Group (www.picmg.org).

PICMG 2.16 The standard from PICMG which uses Ethernet in the backplane for switching traffic between the cards in the PICMG chassis.

PIM Protocol Independent Multicast. A multicast routing protocol specified by the IETF.

ping Packet Internet Groper. A mechanism whereby an ICMP echo request packet (ping request) is sent from one TCP/IP communications device to another to verify its connectivity and status. The receiver of the ping request sends an ICMP echo reply (ping response) to the sender.

PMC PCI Mezzanine Card. A card with a processor and/or other peripherals for adding functionality to an embedded system.

polling A method of programming in which the processor repeatedly checks for a specific event or events.

POSIX   Portable Operating System Interface. A standard environment for enabling the portability of applications software. Originally a work of the Open Group, it has been adopted by the IEEE and ISO.

POTS   Plain Old Telephone Service. Used to describe the "traditional" telephone network.

PPP   Point to Point Protocol. A protocol specified by the IETF for transmission of Layer 2/3/4 packets over a serial link.

predicate   A parameter in a state event table used as an additional qualifier for an action routine.

process   An operating system abstraction that allows different operations to take place concurrently. Memory protection is typically enforced between processes.

product requirements document   A document prepared by Marketing to provide specifications requirements for the product.

protocol   A set of messages and rules for communicating between peer entities. Protocols are usually specified in standard specifications.

protocol stack or protocol suite   The software implementation of a protocol.

pseudo header   A header that is used only in calculating the TCP or UDP checksum.

PVC   Permanent Virtual Circuit. A virtual circuit that is set up manually and stays on.

QNX™   A real-time operating system from QNX Software Systems with a strong base in message passing.

QoS   Quality of Service. Provides guarantee of quality of network.

RED   Random Early Detection. Congestion is detected before it occurs, and packets are discarded randomly to prevent it.

redundancy   An implementation in which a primary resource is backed up by an identical resource for fault tolerance/high-availability purposes.

reentrant code   Code that has been written so that it uses no global or local persistent context across invocations. So, a reentrant function can call itself any number of times with no impact on any variables or states used by the function.

RFC   Request for Comments. Documents specified under the IETF specifying various protocols.

RIP   Routing Information Protocol. A routing protocol to provide routing updates between routers to build a routing table.

route leaking or route redistribution   The process of redistribution of routes between various protocols in a system. For example, BGP can "leak" some routes to OSPF but not others.

router   A device that examines datagram headers to forward them to the next hop destination.

routing information base (RIB)   The complete routing table constructed by a Route Table Manager (RTM) based on routing information from all routing protocols.

RTM   Route Table Manager. A module which deals with route redistribution and construction of a Routing Information Base. It may also construct a Forwarding Information Base (FIB) culled out of the RIB for distribution to the individual forwarding elements.

RTOS   Real Time Operating System.

SDL Specification and Description Language defined by ITU-T for system and protocol design, especially in telecommunications. The presentation is in a graphical form.

select A call used in the socket API to make a process wait on a set of descriptors for events on any of the descriptors.

semaphore A variable which helps support synchronization between processes, so that a process which tries to get a semaphore which has already been taken blocks until the semaphore is released.

SerDes Serializer/Deserializer. An integrated circuit that converts parallel data to the serial form and vice versa.

serial port A slow-speed interface port that transmits data serially.

shared memory A memory that is shared for use by different processes.

signaling A sequence of steps in hardware and/or software to set up desired behavior. For example, SS7 signaling is used to set up an end-to-end voice call through a circuit-switched network.

slow path The path in the processing logic in which less frequent conditions and exceptions are handled. This path usually has several conditional paths.

SNMP Simple Network Management Protocol. A protocol specified by the IETF for management of network devices. The protocol message interchange is between a network manager and the managed device.

socket API The API available via the (originally) UNIX-based abstraction of a connection endpoint.

soft switch A system that handles both the internet and the telephone network and serves as an interface between the two.

spanning tree protocol A protocol specified by the IEEE 802.1D standard. STP is used in a Layer 2 bridged LAN topology to detect loops in the topology.

sparse matrix A matrix with very few non-zero or valid entries. In a SET, this implies a table with very few entries in which the action routine is a valid operation. The other entries are No Ops (no operation or action).

SPF calculation The Dijkstra Shortest Path First (SPF) calculation used for finding the shortest path between a source and destination. It is used by link state protocols like OSPF and IS–IS to build the routing table.

split control plane An implementation in which the control plane functions like routing and signaling protocols are partially implemented on the control card and line card respectively. The two parts of the Split Control Plane interact with each other to provide the appearance of a single control plane. Contrast with Monolithic Control Plane.

SRAM Static Random Access Memory. Faster than Dynamic Random Access memory and is generally used for cache memory.

SS7 Signaling System 7. The protocol used for call setup and forwarding in voice network.

state event table (SET) A table-based representation (and implementation) of a state machine. The most common SET representation consists of M rows and N columns, where each column corresponds to a state and the row corresponds to an event. The whole table appears as a matrix of entries, with each entry representing the Action to be taken on the occurrence of the event in the specific row, and the next State to transition to.

state machine Also called Finite State Machine or FSM. This is a construct used to specify the various states that a protocol can assume, which events are valid for those states, along with the action to be taken on specific events.

stateful A protocol or system which maintains history or state about past events and transactions and uses this history for future transactions. Contrast with Stateless.

stateless A protocol or system which does not depend upon its previous events or transactions for its future behavior. Contrast with Stateful.

static allocation A method of allocating the memory required for data before the start of execution. This is usually done via static definitions of the data in the source code itself.

STREAMS A framework originally specified in AT&T UNIX for building modular communications infrastructure and applications. STREAMS permit the addition and removal of protocol processing modules within the UNIX kernel.

strict layering A situation in which a layer does not use knowledge of its upper or lower layers for its own operation.

SubAgent Defined in the AgentX architecture, a subagent is a module which provides the management interface on a line card. It does not interface directly with the manager but only with the master agent on the control card. The AgentX protocol is used between the control card and the line card.

switch A device which forwards traffic between its interfaces. A Layer 2 Switch is a special case of a bridge. A Layer 3 Switch is a router. Other types of switches include ATM and Frame Relay Switches.

switch fabric Usually a chip or card used to direct traffic between two ports or line cards. A fabric permits multiple traffic streams to be switched in parallel.

synchronous call Another name for a blocking call which does not return until the function completes its operation.

task A thread of execution. Real-time environments frequently use this term to depict code which has its own execution context including program counter, stack and local variables.

TCP Transmission Control Protocol. Most widely used transport protocol for reliability.

telnet A protocol used for terminal access to hosts in the Internet. Telnet runs over TCP.

thread A schedulable entity that can share data with other such entities.

time division multiplexing (TDM) A method of sharing a medium using time slots in which data from multiple senders is sent over multiple time slots.

timer A function required to keep track of the temporal activities required by the communications system. Timers can be one shot (they only fire once) or continuous. One-shot timers are required for actions that are performed only once after a certain period of time. Continuous timers start off with a timeout value and count down from there. On reaching zero, they generate an event and reset the time-out value to count down again.

timer block   A data structure which stores the countdown value for a timer and the parameters related to it. These can include the ID of the task/module which started the timer along with the routine to be called upon timeout (with its parameters). Timer blocks are linked to each other to form one or more timer lists.

timer management   The mechanism to handle multiple timers for one or more protocols or the entire system. This mechanism involves starting and stopping timers as well as signaling tasks and modules with the appropriate parameters upon a timeout.

timer management task (TMT)   A task which manages the timers for all the tasks in the system. This is the only task which maintains the lists of timer blocks.

TL1   Transaction Language 1. A standard originally specified by Bellcore (now Telcordia) for the management of network elements. It is a set of ASCII-based instructions or messages used by a network manager (also called an Operations Support System or OSS in telecom parlance) to manage a network element or device.

TLV   Type Length Value. A scheme of encoding of messages where the first field represents the type of parameter, the second represents the length of the parameter, and the third represents the actual parameter value. Each of the fields could span multiple bytes.

TOE   TCP Offload Engine. A hardware device or board which is used to improve the performance of a system by terminating and processing TCP connections. This frees up the CPU to handle other tasks.

traffic engineering   An aspect of network engineering concerned with the performance optimization of traffic on the network. The most common is the optimization with respect to over-utilization and under-utilization of paths and links in the network.

trap   A notification or alert sent by an SNMP agent to one or more SNMP managers using an SNMP Trap PDU. The notification can be the result of an alert sent by a protocol or system task to the SNMP Agent task.

trie   A tree structure for storing strings with a common node for each prefix. Often used as a structure for storing IP network address prefixes for faster searches. The name comes from re*trie*val and is pronounced "tree."

UDP   User Datagram protocol. A connection-less transport protocol less reliable than TCP.

unicast address   The network address of a specific station.

UNIX   An operating system originally developed at Bell Labs and commercialized by AT&T. Another version was developed in the late 70s at the University of California, Berkeley. This was known as Berkeley Systems Distribution (BSD) UNIX.

user mode   The mode in UNIX where applications operate as individual processes with memory protection enforced between processes, as well as between the process and the operating system kernel.

VLAN   Virtual Local Area Network. A logical partition of a Layer 2 network in which the end nodes may be part of the same physical LAN but belong to different VLANs.

VxWorks™ The name of a real-time operating system product from Wind River Systems.

WAN Wide Area Network. A network which spans a large geographic area. Frame Relay and X.25 are examples of WAN technologies.

warm standby A redundancy scheme in which the standby card was initialized but still requires operator intervention and/or last configuration update for the primary before it can take over.

Web-based management HTTP-based management in which the user connects with the communications device via a browser and configures/monitors parameters on the device.

WRED Weighted Random Early Detection. A variation of RED which uses weighted probabilities for the priorities of packets to select the packets to discard.

XML Extensible Markup Language. A "meta language" used to describe other languages, it is used to improve the functionality of the Web as well as provide a uniform language for communicating between applications on different hosts.

References

1. Accelerated Technologies, Nucleus Real Time Operating System, http://www.atinucleus.com/embedded/nucleus.html

2. AT&T–Unix System V Release 4: *Programmer's Guide: Streams* (AT&T Unix System V, Release 4. System Programmer's Series)

3. Berezin, Tanya. "Writing a Software Requirements Document," http://www.sims.berkeley.edu/courses/is208/s02/ReqsDoc.pdf

4. Binstock, Andrew. "Hashing Rehashed," Dr. Dobb's Journal, April 1996

5. Broadcom Corporation Switching Product Family, http://www.broadcom.com/entnetstrata.html

6. Brooks, Frederick P. *The Mythical Man-Month: Essays on Software Engineering*, Addison Wesley, April 1995

7. Cisco Systems 12000 Series Routers, http://www.cisco.com/en/US/products/hw/routers/ps167/index.html

8. Cisco Systems. "Saving and Restoring Configurations on IPX, IGX, and BPX Nodes:" http://www.cisco.com/warp/public/74/110.html

9. Comer D. *Computer Networks and Internets*, 3rd Edition, Prentice Hall, 2001

10. Comer, Douglas. *Network Systems Design Using Network Processors*, Pearson Prentice Hall, January 2003

11. Data Connection Limited "High Availability Framework," http://www.dataconnection.com/mpls/highavfr.htm

12. Davie, B. and Rekhter, Y. *MPLS Technology and Applications*, Morgan Kaufmann, 2000

13. Dijkstra, E.W. "A Note on Two Problems in Connection with Graphs," Numer. Math., October 1959

14. Donald R. Morrison. "PATRICIA—Practical Algorithm To Retrieve Information Coded in Alphanumeric," ACM Journal, Vol. 15, No. 4, October 1968, pp. 514–534

15. Ganssle, Jack. "The Seven Habits of Highly Defective Developers," Embedded Systems Programming, July 1998

16. Ganssle, Jack G. "Interrupt Latency," Embedded.com, 1 October 2001

17. Ganssle, Jack G. "Introduction to Reentrancy," Embedded.com, 15 March 2001

18. Ganssle, Jack G. "The Challenges of Real Time Programming," Embedded Systems Programming, July 1998

19. Ganssle, Jack G. "The Art of Designing Embedded Systems," Newnes, 1999

20. Goralski, W.J. *Introduction to ATM Networking*, McGraw-Hill, 1995

21. Halabi, Sam. *Internet Routing Architectures*, Second Edition, Cisco Press, 2001

22. Hawley, Greg. "Selecting a Real Time Operating System," Embedded Systems Programming, March 1999

23. Huitema, Christian. *Routing in the Internet*, 2nd edition, Prentice-Hall, 2000

24. Husak, David. "Network Processors—A Definition and Comparison"—White Paper, 2000. http://e-www.motorola.com/collateral/M957198397651.pdfhttp://e-www.motorola.com/collateral/M957198397651.pdf

25. IEEE Standard 802.1D. "MAC Bridges," 1998

26. IEEE Standard 802.1Q. "Virtual Bridge Local Area Networks," 1998

27. Ixia Corporation ANVL Suite, http://www.ixiacom.com/products/caa/

28. Jain, Raj , "A Comparison of Hashing Schemes for Address Lookup in Computer Networks," IEEE Transactions on Communications, Vol. 40, No. 3, October 1992, pp. 1570–1573, http://www.cis.ohio-state.edu/~jain/papers/hash_iee.htm

29. Johnson, E. and Kunze A. *IXP1200 Programming: The Microengine Coding Guide for the Intel IXP1200 Network Processor Family*, Intel Press, 2001

30. Jones, Anthony and Ohlund, Jim. *Network Programming for Microsoft Windows*, Microsoft Press, August 1999

31. Juniper Networks M-Series Routers http://www.juniper.net/products/ip_infrastructure/m_series/index.html

32. Keshav S. and Sharma, R. "Issues and Trends in Router Design," IEEE Communications Magazine, May 1998

33. Keshav, S. *An Engineering Approach to Computer Networking*, Addison Wesley, 1997

34. Labrosse, Jean "MicroC OS II: The Real Time Kernel", CMP Books, 2002.

35. Martin, Robert C. "UML Tutorial: Finite State Machines," Engineering Notebook Column, C++ Report, June 1998. Available at http://www.objectmentor.com/resources/articles/umlfsm.pdf

36. Marvell Semiconductor Switching Product Family, http://www.marvell.com/products/switching/index.jsp

37. McKusik, Marshall Kirk, et al. *The Design and Implementation of the 4.4BSD Operating System*, Addison Wesley, April 1996

38. MontaVista Software. "Embedded Linux—Ready for Real-Time," White Paper, 2001. http://www.mvista.com/dswp/RTReady.pdf

39. Murphy, Niall. "Watchdog Timers," Embedded Systems Programming, November 2000

40. Nework Processing Forum www.npforum.org

41. Newton, Harry. *Newton's Telecom Dictionary—18th Updated and Expanded Edition*, CMP Books, March 2002

42. Nix, David. "Common Architectures for Communications," Embedded Systems Programming, November 1999

43. Orr, Michael. "When Network Design Meets Chaos Theory," Communications Systems Design, February 2003

44. OSE Systems, OSE Real Time Kernel, http://www.ose.com/prodserv/coreos/

45. Perlman, Radia. *Interconnections: Bridges, Routers, Switches, and Internetworking Protocols*, 2nd edition, Addison-Wesley, 1999

46. Peterson, L. and Davie, B. *Computer Networks—A Systems Approach*, Morgan Kaufmann, 2000

47. QNX Software Systems, QNX Neutrino Operating System, http://www.qnx.com/products/ps_neutrino/

48. RFC 1058. "Routing Information Protocol," June 1988

49. RFC 1573. "Evolution of the Interfaces Group of MIB-II"

50. RFC 1661. "The PPP Protocol," July 1994

51. RFC 1771. "A Border Gateway Protocol (BGP-4)," March 1995

52. RFC 1812. "Requirements for IP Routers," June 1995

53. RFC 2309. "Recommendations on Queue Management and Congestion Avoidance in the Internet"

54. RFC 2328. "OSPF Version 2," April 1998

55. RFC 2401. "Security Considerations for the Internet Protocol," November 1998

56. RFC 2741. "Agent Extensibility (AgentX) Protocol Version 1," January 2000

57. Ruiz-Sanchez, et al., "Survey and Taxonomy of IPAddress Lookup Algorithms," IEEE Network, March/April 2001

58. SDL Forum Society, "What is SDL?" http://www.sdl-forum.org/SDL/index.htm

59. Seifert, Rich. *Gigabit Ethernet: Technology and Applications for High-Speed LANs*, Addison Wesley, April 1998

60. Seifert, Rich. *The Switch Book: The Complete Guide to LAN Switching Technology*, John Wiley & Sons, 2000

61. Seifert, Rich. *Gigabit Ethernet*, Addison Wesley, 1998

62. Service Availability Forum www.saforum.org

63. Simon, David E., "An Embedded Software Primer", Addison-Wesley, 1999.

64. Sridhar, T. "Control and Data Plane Issues in Communications Software," Communications Design Conference, September 2002

65. Sridhar, T. "Reentrancy in Protocol Stacks," Embedded Systems Programming, November 2001

66. Sridhar, T. "Tackling Multiboard Networking Designs," Commsdesign.com, April 6, 2001

67. Sridhar, T. and Srinivasan, Manikantan. "Modules ease programming task," EE Times, November 4, 2002

68. Sridhar, Thayumanavan. "Layer 2 and Layer 3 Switch Evolution," Cisco IP Journal, September 1998

69. Sridhar, Thayumanavan. "Layer 3 Switch Design," Communications Systems Design, April 1998

70. Sridhar, Thayumanavan. "Strategies for Communications Systems Software Design," Embedded Systems Programming, June 1998

71. Stallings, William. *Data and Computer Communication*, 6th edition, Prentice-Hall, 1999

72. Stallings, William. *SNMP, SNMP v2, SNMPv 3, and RMON 1 and 2*, Addison-Wesley, December 1998

73. Stallings, William. *SNMP, SNMPv2 and CMIP*, Addison-Wesley, 1993

74. Stevens, Richard. *Unix Network Programming*, 2nd edition, Prentice Hall, 1998

75. Stewart, Dave. "30 Pitfalls for Real Time Software Developers," Embedded Systems Programming, October and November 1999

76. Stewart, Dave. "Introduction to Real Time," Embedded Systems Programming, November 2001

77. Tanenbaum, Andrew S. *Computer Networks*, 4th edition, Prentice Hall, 2002

78. Tennies, Nathan. "Software Matters for Power Consumption," Embedded Systems Programming, February 2003

79. TICS. Tutorial on Timer Management, http://www.cris.com/~Tics/tics0197b.htm

80. TL1.com. "Beginners Guide to TL1," http://www.tl1.com/library/TL1/Overview/Beginners_Guide_to_TL1.html

81. Washington University, St. Louis, "The Adaptive Communications Environment (ACE)," http://www.cs.wustl.edu/~schmidt/ACE.html

82. Wind River Systems, VxWorks Developers Toolkit, http://www.windriver.com/markets/platformvdt/index.html

Index

Page numbers in *italics* indicate terms in the Glossary.

Printed and bound by CPI Group (UK) Ltd, Croydon, CR0 4YY

24/10/2024

01778606-0001